MATHEMATICAL MODELS IN BIOLOGY AND MEDICINE

MATHEMATICAL MODELS IN BIOLOGY AND MEDICINE

Proceedings of the IFIP-TC4 Working Conference on
Mathematical Models in Biology and Medicine,
Varna, 6–11 September 1972

Edited by

N. T. J. BAILEY
World Health Organization, Geneva

Bl. SENDOV
Mathematical Institute, Bulgarian Academy of Sciences, Sofia

R. TSANEV
Biochemical Institute, Bulgarian Academy of Sciences, Sofia

1974

NORTH-HOLLAND PUBLISHING COMPANY – AMSTERDAM · LONDON
AMERICAN ELSEVIER PUBLISHING COMPANY, INC. – NEW YORK

© IFIP-1974

All rights reserved. No part of this publication may be reproduced, stored in a retrieval system, or transmitted, in any form or by any means, electronic, mechanical, photocopying, recording or otherwise, without the prior permission of the copyright owner.

Library of Congress Catalog Card Number: 73-81533
North-Holland ISBN: 0 7204 2087 3
American Elsevier ISBN: 0 444 10533 6

Published by:
North-Holland Publishing Company–Amsterdam
North-Holland Publishing Company, Ltd.–London

Sole distributors for the U.S.A. and Canada:
American Elsevier Publishing Company, Inc.
52 Vanderbilt Avenue
New York, N.Y. 10017

Printed in the Netherlands

PREFACE

Mathematical modelling is a universal method for studying nature. Recently, this method has been widely applied to various fields of knowledge. Its application to biology and medicine appears to be extremely successful. The IFIP TC-4 organized a Working Conference on Mathematical Models in Biology and Medicine in answer to the continuously increasing interest in mathematical modelling in all countries. The conference was held in Varna, Bulgaria, on September 6–11, 1972 as a satellite event of the Third Congress of the Bulgarian Mathematicians.

The following problems were discussed:

(a) genetic control mechanisms and their role in the study of differentiation, cell-virus and carcinogenesis;

(b) biochemical control mechanisms and their role in cellular and physiological regulation;

(c) the kinetics of compartment models and their role in pharmacokinetics;

(d) physiological control mechanisms;

(e) populational and ecological control mechanisms;

(f) information theory in biology and medicine;

(g) medico-social control mechanisms.

Participation was restricted to about forty invited speakers from 12 countries. Nineteen papers were presented and discussed at the conference. A panel session was also held. Edited transcripts of the discussions and the panel session are included in this volume as well.

The Organizing Committee for the conference consisted of:

Prof. Dr. Bl. Sendov – chairman,
Prof. R. Tsanev – vice-chairman,
Dr. D. Dojchinov,
E. Kolarova – secretary.

The financial support of the conference was ensured by IFIP and the Bulgarian Academy of Sciences.

Special thanks are due to IFIP and the Bulgarian Academy of Sciences who much helped in the organization of the conference ensuring its financial support.

CONTENTS

Preface   v

List of contributors   ix

Mathematico–statistical analysis of the cyclic recurrence in the droplet infections of communicable diseases
    S. Bachev and E. Petkova   1

Medico-social control mechanisms
    N. T. J. Bailey   9

On reconstructing recessive information in homogeneous biological systems
    G. Hempel   13

Mathematical models in cell cycle kinetics
    B. Jansson   21

Evolution and control of a biochemical system
    J. P. Kernevez   41

Reflections on some algorithms for the solution of the inverse problem (identification or adjustment) for linear compartmental models
    C. Monot and J. Martin   49

Resistance of crops to diseases: a game-theory model
    J. Pešek   71

The information content of kinetic data
    J. G. Reich and I. Zinke   77

A mathematical model of cellular differentiation
    Bl. Sendov and R. Tsanev   81

A mathematical scheme for morphogenesis: structural stability and catastrophes
    R. Thom   93

Models of the dependence of hospital utilization upon medico-social factors
    I. Väänänen, G. Bäckman, A. S. Härö, J. Perälä and O. Vauhkonen   97

A method of computer simulation used to study some sublethal effects of irradiation on cell kinetics
    A.-J. Valleron   111

PANEL DISCUSSION

Qualitative and quantitative models
    R. Thom   135
Adaptive model building and computer assisted analysis of biomedical data
    T. Groth   137
Modelling
    N. T. J. Bailey   141
Pragmatic justification of the "model" in medicine
    J. Martin and C. Monot   143

Author index   151

LIST OF CONTRIBUTORS

S. Bachev, Research Institute of Social Hygiene and Public Health, D. Nesetirovstreet 15, Sofia, Bulgaria

G. Bäckman, Children's Castle Hospital, Lastenlinnantie 2, 25 Helsinki, Finland

N. T. J. Bailey, World Health Organization, 1211 Genève 27, Switzerland

T. Groth, Uppsala Data Centre, University of Uppsala, Box 2103, 75002 Uppsala, Sweden

A. S. Härö, Children's Castle Hospital, Lastenlinnantie 2, 25 Helsinki, Finland

G. Hempel, Sektion Forstwirtschaft, Technische Universität Dresden, Pienner Strasse 8, 8223 Tharandt, D.D.R.

B. Jansson, National Large Bowel Cancer Project, M. D. Anderson Hospital and Tumor Institute, Houston, Texas 77025, U.S.A.

J. P. Kernevez, Faculté des Sciences, 6 bd Gabriel, 21000 Dijon, France

J. Martin, Section d'Informatique Médicale, U.E.R. de Médicine, Université de Nancy, 30 rue Lionnois, 54 Nancy, France

C. Monot, Centre National de la Recherche Scientifique, 54 Nancy, France

J. Perälä, Children's Castle Hospital, Lastenlinnantie 2, 25 Helsinki, Finland

J. Pešek, Department of Biometrics, Academy of Agriculture, Hrušovany u Brna, Czechoslovakia

E. Petkova, Research Institute of Social Hygiene and Public Health, D. Nesetirovstreet 15, Sofia, Bulgaria

J. G. Reich, Zentralinstitut für Molekularbiologie, Bereich Methodik und Theorie, Akademie der Wissenschaften der D.D.R., Lindenberger Weg 70, 1115 Berlin-Buch, D.D.R.

Bl. Sendov, Mathematical Faculty, University of Sofia, ul. A. Ivanov 5, Sofia 26, Bulgaria

R. Thom, Institut des Hautes Etudes Scientifiques, 35 Route des Chartres, 91 Bures-sur-Yvette, France

R. Tsanev, Biochemical Institute, Bulgarian Academy of Sciences, 13 Sofia, Bulgaria

I. Väänänen, Children's Castle Hospital, Lastenlinnantie 2, 25 Helsinki, Finland

A.-J. Valleron, Institut National de la Santé et de la Recherche Médicale, Unité de Recherches Statistiques, 16bis avenue Paul-Vaillant Couturier, 94800 Villejuif, France

O. Vauhkonen, Children's Castle Hospital, Lastenlinnantie 2, 25 Helsinki, Finland

I. Zinke, Zentralinstitut für Molekularbiologie, Bereich Methodik und Theorie, Akademie der Wissenschaften der D.D.R., Lindenberger Weg 70, 1115 Berlin-Buch, D.D.R.

MATHEMATICO–STATISTICAL ANALYSIS OF THE CYCLIC RECURRENCE IN THE DROPLET INFECTIONS OF COMMUNICABLE DISEASES

S. BACHEV and E. PETKOVA

Research Institute of Social Hygiene and Public Health, Sofia, Bulgaria

1. Introduction

The cyclic fluctuations in tracing up the dynamics of the droplet infections of communicable diseases is an object of epidemiologic investigation. However, the scientific as well as the practical investigations carried out usually lead to an elementary description of the phenomenon. The cause of the above-mentioned phenomenon may be attributed to the lack of an appropriate mathematico–statistical method for quantitative analysis of the "cyclicity" in phenomena with a dynamic course.

The term "cyclicity" as defined in the epidemiology signifies a series of epidemic explosions, occurring at a definite period of time.

The authors have set themselves the task to build up a method providing a possibility to carry out a profound statistical–epidemiological analysis of the cyclicity in the course of the droplet infections, as a part of the analysis relevant to the dynamic changes occurring in them.

The concrete problems to be considered are as follows:

(1) the determination of the algorithm, proving the regular cyclicity course in the dynamics of the droplet infections;

(2) the choice of a suitable mathematical apparatus for the description and demonstration of cyclicity;

(3) the determination of the character, the empirical description and the cognitive qualities of the indices characterizing the cycle.

2. A proof of cyclicity with a regular course

The studies carried out revealed as suitable for this purpose two of the well-known mathematico–statistical methods, namely, the range analysis (non-parametric method for calculating the correlation coefficient) and the dispersion analysis. The latter method appeared to be the more convenient for the following reasons:

(1) The non-parametric methods are less sensitive and reflect the phenomena analysed less accurately.

(2) The range analysis is not in a position to determine and differentiate the combined effect of the long acting cyclic and noncyclic factors, which are reflected in the dynamics of the phenomena.

The dispersion analysis is used in its monofactorial modification, with factor "cycle". The successful application of the dispersion analysis requires a certain analytical grouping of the data of the statistical dynamics determining the cyclicity. The analytical grouping is brought to limiting the uprise periods (peak of the epidemic wave), the depression periods (the minimum of the epidemic wave), and intermediate periods (data of the dynamic series of the two moments of the epidemiological wave having extreme values). With this view of a more precise presentation of the intensity of the morbidity, i.e. eliminating the effect of the qualitative changes in the environment where the disease is manifested – the sensitive contingents of the population – it is necessary to work with relative values of the intensity, like the morbidity rate per hundred thousand of the population.

The results obtained by means of the dispersion analysis point to the presence of dynamic cycles with a regular course in the case of measles, diphtheria, pertussis and scarlet fever ($P < 0.01$).

The above results give sound grounds for the existence of such a manner of interpretation and analysis.

3. Description and analysis of the cycle

A tentative differentiation can be made between proving and analysing cyclicity or cyclic recurrence. In either case it is a question of analysis in the general sense of the word. But in the case which in practical work is designated as analysis we primarily use indices which give an analytical description of the cycle. These indices include the width of the cycle, the amplitude of variation of the cycle, and the indices of the cycle.

3.1. *Mean width of the cycle*

This is the mean length of time between two epidemic peaks or between two epidemic minima, and is calculated according to the formula

$$C_w = (x_n - x_0)/m, \tag{3.1}$$

where
- C_w = width of the cycle,
- x_n = calendar year of the last rise of the observed period,
- x_0 = calendar year of the first rise of the observed period,
- m = number of observed cycles.

Substituting the pertussis data (Fig. 1) in the above formula, we have

$$C_w = (1970 - 1954)/4 = 4 \text{ years.}$$

The interval of confidence of the mean value obtained above is calculated in accordance with the conventional method, the elements of the variational series being formed by the lengths of the separate cycles in years.

With a view to a more complete characterization of the cycle, the following two-width indices may be used: the mean width of the semiperiod of depression (the time from the maximum to the minimum of the epidemic wave), and the mean width of the semiperiod of the rise (the time from the minimum to the maximum of the epidemic wave). Presenting the width index in greater detail provides additional information with regard to the analysis of the drop and the rise of the morbidity.

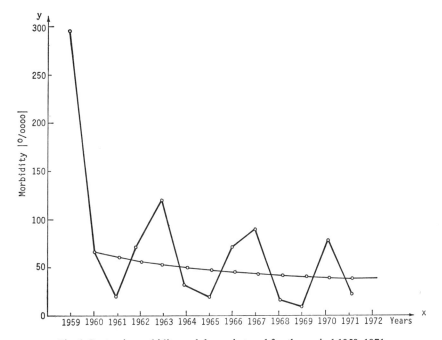

Fig. 1. Pertussis morbidity and dynamic trend for the period 1959–1971.

3.2. The amplitude of variation

This represents the difference between the minimum and the maximum value of the epidemic wave. The index relevant to the mean amplitude of variance is presented by the formula

$$R_c = p^{-1} \sum_{i=1}^{p} x_{i\,max} \Big/ q^{-1} \sum_{i=1}^{q} x_{i\,min}, \quad (3.2)$$

where

R_c = mean amplitude of the variance,
$x_{i\,max}$ = morbidity at the maximum of the epidemic wave,
$x_{i\,min}$ = morbidity at the minimum of the epidemic wave,
p = number of maxima of the epidemic waves;
q = number of minima of the epidemic waves.

With the pertussis model the value of this index is 406.5%, which shows that the morbidity rate at the minimum of the epidemic wave is about 4 times lower than at the maximum.

3.3. Indices of the cycle

These are indices quantitatively describing the elements (maximum or peak, minimum, period of rise and depression period) of the cycle.

The indices of the cycle are calculated with the chain or serial application of a number of methods. The need for the use of the respective calculating methods is connected with the dynamic statistical series. As the authors have some differences in view with regard to the generally accepted definitions of the dynamic series, they treat only some parts of the theory.

It is generally accepted that a given dynamic series, that is its values, is determined by the joint action of two groups of factors, namely those with a long action and those with incidental action. The former determine the regularities in the changes in time, while the latter are the fluctuations of the data in the dynamic series. By means of the so-called methods for the elimination of the permanent trend, we can consider the effects of the so-called casual factors, which reflect only the effect of the permanently active factors. Our studies showed that by using the above methods the effects of such permanently active factors, which affect and lead to systematic fluctuations, such as those appearing in the cyclic ones, are also eliminated. Consequently, if we take the ratio of the actual values of the dynamic series and those of the equalized one tracing the permanent trend, we obtain a new dynamic series reflecting the effect of both the casual and the cyclic factors. A possibility for demonstrating the cyclicity phenomenon is provided by the use of the averaging procedure as a method for eliminating the effect of the casual or incidental factors.

The essential part of the calculation process, reflecting the above-mentioned state, may be presented in the following three stages:

(1) calculation of the permanent trend of the dynamic series;

(2) reducing to percent ratios with the equalized values to the corresponding actual values;

(3) representing as mean values the obtained results for the corresponding parts of the epidemic cycle – cyclicity indices.

On the basis of the pertussis data, the above-mentioned three stages can be specified as follows:

(1) The permanent trend is traced by the equation

$$\log y = 2.46188 - 0.68158 \log x. \tag{3.3}$$

(2) The cyclicity indices (I_c) are calculated as a ratio of the actual values with regard to the equalized values of the dynamic series and are shown in Table 1.

Table 1

Year	Code (x)	Morbidity (y)	Trend (y_t)	Index (I_c)
1959	8	290.3	70.2	413.53
1960	9	66.3	64.8	102.31
1961	10	18.2	60.3	30.18
1962	11	76.0	56.5	134.51
1963	12	119.9	53.3	224.95
1964	13	28.3	50.4	56.15
1965	14	18.1	47.9	37.79
1966	15	69.2	45.7	151.42
1967	16	87.9	43.8	200.68
1968	17	15.1	42.0	35.95
1969	18	9.0	40.4	22.28
1970	19	77.3	38.9	198.71
1971	20	2.4	37.6	56.91

(3) Presenting the indices as mean values, with a view to eliminating the incidental fluctuations, the elements of the epidemic cycle gave the following dimensions:

$\bar{I}_{max} = 259.47\%$, $\bar{I}_{min} = 30.08\%$,
$\bar{I}_{rise} = 142.97\%$, $\bar{I}_{depression} = 62.83\%$.

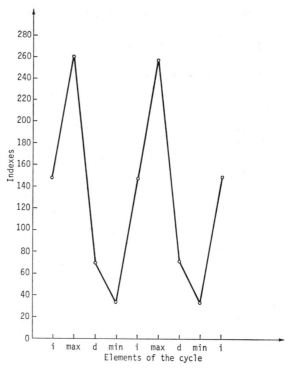

Fig. 2. Indices of the pertussis cycle.

The epidemic cycle on the basis of the calculated mean indices is presented in the following manner (Fig. 2):

The conventional approach to the prognosis of the infectious diseases is to extrapolate the data relevant to the dynamic series on the basis of the calculated permanent trends. In these cases the calculated values contain a systematic error, due to not taking into account the cyclicity effect. This error can be avoided to a certain extent by correcting the extrapolated values for the cyclicity indices. The calculation is presented by the formula

$$y_{ip} = y_{tp} \cdot \bar{I}_j, \tag{3.4}$$

where

y_{tp} = extrapolated value of the permanent trend (prognosis of the morbidity),

$\bar{I}_j$ = mean index of the cycle element, where j may be the maximum, the minimum, the rise or the depression,

y_{ip} = corrected value of the prognosis with the cyclicity effect.

Prognosticating the pertussis morbidity in the above conditions, the extrapolated values for the 1972–1976 period were calculated and corrected in accordance with the effect of the epidemic cycle, as shown in Table 2 and Fig. 3.

Table 2

Year	Code (x)	Trend (y_t)	Prognosis (y_p)
1972	21	36.4	10.94
1973	22	35.2	50.33
1974	23	34.2	88.74
1975	24	33.2	20.85
1976	25	32.3	9.72

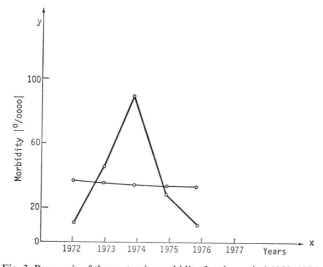

Fig. 3. Prognosis of the pertussis morbidity for the period 1972–1976.

The prognosis of the droplet infections of communicable diseases is plausible, provided that the factors determining the permanent trend and cyclicity of the diseases do not change their effect in the prognosticated period.

As a conclusion it may be stated that on the basis of the described methodology we can obtain a comparatively good proof, description, analysis and prognosis of the cyclicity in the epidemic process of the droplet infections by making use of the dynamic statistical series.

MEDICO-SOCIAL CONTROL MECHANISMS

N. T. J. BAILEY

World Health Organization, Geneva, Switzerland

One aspect of medico-social control that needs much closer attention and development is the use of systems models to understand and control the planning, provision and maintenance of health services. Moreover, it is generally recognized that improvement in health is achieved not only through the curative and preventive services of the health sector as such, but is also strongly influenced by developments and activities in other sectors such as economics, transportation, education, agriculture, etc.

In most realistic planning situations, whether at micro- or macro-levels, it is usually difficult to define a convenient objective function to be maximized, and we inevitably become involved in difficult problems of trade-offs between non-commensurate quantities. Thus in planning an appointment system for an out-patient clinic, the time waited by the patients is inevitably related to the time wasted by the consultants, but the qualities of patients' time and consultant's time may be difficult to compare except in very broad terms. Again, at the other end of the scale, we may be trying to improve the health of a rural area in a developing country, where finance and manpower are severely limited. There may be awkward problems in striking a balance between curative and preventive services, between several different diseases that should ideally be eliminated or at least held down to some acceptable level, and between providing health care to different age and sex groups or other social sections of the community. And at this level improvements in agriculture, transportation or education may have more effect on health status than a more direct expenditure on disease control as such.

A common approach nowadays to such problems is to attempt to define a series of alternative strategies available to the decision maker, each strategy having an associated scenario describing the probable broad outcomes of that particular choice. The decision maker does not attempt to construct and maximize a single objective function, but chooses subjectively between the broad alternatives. There are good reasons for believing that such a subjective choice, based on well-presented, explicit alternatives using predictions derived from carefully developed systems models, is far better than the more traditional approach of pure intuitive judgment using only limited data and unsupported speculation.

However, if we adopt this approach, major questions arise of how such systems models are to be developed and handled, what degree of complexity is desirable for realism, what degree of simplicity is acceptable for ease of handling and interpretation, what aspects of the models can be tested and evaluated, what features are essentially exercises in a logical analysis of assumptions and what features have the status of scientific hypotheses to be examined, what conclusions from any model are sensitive to changes or errors in some of the assumptions, etc.

Such questions are rarely worth examining in relation to a purely theoretical model, though they may well be worth studying in a general model that can be seen to have components that are easily related to typical aspects of the real world. Thus a general model of a health care system might take into account changes in population structure, including mortality, morbidity or disability, a range of alternative health care and medical services, together with the possibility of modifying the various transition rates or probabilities by health education, improvements in transportation, etc. This could well reveal certain prominent features of the system that provided general insight into the kind of relationships to be expected between certain control measures and the expected outcomes, such as improvements in efficiency or health status. It might also indicate the existence of counterintuitive phenomena, that would need further investigation.

Without some empirical content such results could always be severely criticized on grounds of being totally unrealistic. It would be natural to construct a model that not only had some reasonably typical structure, but also employed parametric values of a kind found in practice. Such models can be found in the literature. They certainly excite greater interest, but are still open to the suspicion that a concrete situation, rich in its own particular complexities, would behave differently.

More convincing in many ways are the attempts to model a specific situation, such as the provision of health services at the regional or national level in a given country, or even the interaction of many sectors, involving health, population, natural resources, environmental pollution, etc., on a world level, as in the MIT project sponsored by the Club of Rome.

However, highly aggregated models of this sort, such as those developed by J. Forrester and D. Meadows for the study of urban and world dynamics, raise a number of crucial issues. Predictions of disasters involving overpopulation, depletion of environmental resources, and steeply rising pollution, are so consistent with common-sense observations that they cannot reasonably be ignored. Yet it is easy for some experts, in say agricultural production or air pollution, to claim with justification that the modelling of the subsystems is simplified to a high degree, and that this simplification has been taken to a point where results are totally unreliable. Other experts, on the other hand, maintain that essential basic features have been retained, and that conclusions are sufficiently accurate to warrant them being taken seriously and to justify special measures of intervention and control.

In this context it seems essential that more effort be put into evaluating the reliability of such systems models. It is quite usual for the constructors of these models to vary some of the parameters and to show that the broad character of certain conclusions is only weakly affected by such changes. When this can be done, it may justify the use of soft data or inspired guesses to fill in gaps in knowledge. On the other hand, the number of parameters that is feasible to examine is severely limited. Even more serious is that the number of alternative formulations of the *structure* of the system, including the degree of complexity entailed, is virtually infinite. While it is accepted that a wide range of phenomena in biology, for example, can be studied in terms of the laws and mechanisms operating on the biological level, without reference to the finer structure involved at chemical or physical levels, little seems to be known about the extent to which aggregated systems models can be reliably operated independently of processes taking place at a more disaggregated level.

Finally, it should be emphasized that the purpose of this note has been not to propose solutions of these problems but to draw attention to sensitive areas urgently requiring more research and elucidation and to stimulate further technical discussion.

ON RECONSTRUCTING RECESSIVE INFORMATION IN HOMOGENEOUS BIOLOGICAL SYSTEMS

G. HEMPEL

Sektion Forstwirtschaft, Technische Universität Dresden, Tharandt, D.D.R.

1. Introduction

Biological populations (for example, plant populations, animal populations, or other organic populations) are usually described respectively by aggregated values or by expectation values of any distinctive marks or qualities ("Merkmale"). A biological population is a dynamic system. A special class of dynamic systems are the homogeneous systems.

Homogeneous systems are systems consisting of a large number of essentially nearly identical active elements with irrelevant individual behaviour for the whole system. It is possible to consider biological populations as homogeneous systems because they also consist of a large number of individuals. The individuals are the elements of the biological system.

For many questions in biological research (for example, medical and agricultural research) not only the aggregated values (they may be called macroscopic values) are interesting but also the distribution of the single quantities (they may be called microscopic values) of the marks. It is possible to obtain the information about the distribution of the microscopic values by the statistical theory of complex systems given by Voss [2, 3] for physical problems. With some extensions and modifications this method has been used in plant populations.

2. The mathematical formulation

The state of an element E (and of the system too) is given by the inputs X and outputs Y and the interior states Z. This can be illustrated as follows:

$$X = \{x\} \to \boxed{E} \to Y = \{y\}.$$
$$Z = \{z\}$$

Every element E is completely described by three sets and two relations between the sets:

$$E = (X, S, Z, R, Y),$$
$$S \subseteq X \times Z \times Z, \quad R \subseteq X \times Z \times Y. \tag{1}$$

So the microscopic state may be defined by $X = \{x\}$, and the individual representation may be denoted by

$$\{x_1, x_2, \ldots, x_n\} = x. \tag{2}$$

The microscopic state is considered as a point in an n-dimensional space of which the coordinates correspond to the states of the single elements. If the sets X and Y are measurable, the space is a metric space.

Since it is only possible to give a probability assertion on the whole set of biological values, in the space we do not use the single points but a point density, i.e.

$$f_n(x) = f_n(x_1, x_2, \ldots, x_n). \tag{3}$$

Then the values are dispersed around the point x in the interval $dx = dx_1 \cdot dx_2 \cdot \ldots \cdot dx_n$ with probability

$$w = f_n(x)\,dx. \tag{4}$$

Because of the probability interpretation of $f_n(x)$ we have

$$0 \leq f_n(x), \quad \int f_n(x)\,dx = 1. \tag{5}$$

Thus $f_n(x)$ is the probability distribution function of the statistical system state with the density of the element states normed to one. In homogeneous systems, aggregated values or mean values are of essential interest for the characterization of the state of the system. We consider therefore

$$A = \sum_{i=1}^{n} a(x_i), \quad \bar{a} = \frac{1}{n}\sum_{i=1}^{n} a(x_i), \quad \bar{a} = \frac{A}{n}. \tag{6}$$

If the state of the system is given by $f_n(x)$, the macroscopic values must be represented by the expectation values given by

$$\bar{a} = \int \left(\frac{1}{n}\sum_{i=1}^{n} a(x_i)\right) f_n(x)\,dx. \tag{7}$$

Since all elements are equivalent, $f_n(x)$ must be symmetrical under an exchange of the x_i. Therefore we have, by changing the names of the single values,

$$\bar{a} = \int a(x_1) f_1(x_1)\, dx_1, \tag{8a}$$

$$f_1(x_1) = \int f_n(x_2, \ldots, x_n)\, dx_2\, dx_3 \ldots dx_n,$$
$$\int f_1(x_1)\, dx_1 = 1, \qquad f_1(x_1) \geq 0; \tag{8b}$$

where $f_1(x_1)$ is the so-called one-point distribution.

The macroscopic values contain relevant information on the system. The aggregation starts with the microrelations

$$a_i = a(x_i), \tag{9}$$

and goes on to the macrorelations

$$\bar{a} = \tilde{a}(\bar{x}) \tag{10}$$

for the relation between the mean values.

Now we need a selection principle for the unique determination of the state of the statistical system from the infinite number of functions $f_n(x)$. This is possible by determining the maximum of the information entropy

$$H^* = -\int f_n(x) \ln f_n(x)\, dx, \tag{11}$$

where H^* is the generalization for continuous variables of the Boltzmann model $H = -\sum_i p_i \ln p_i$ for discrete variables. The logarithm is a monotonous function; therefore the maximum of the information-theoretic entropy corresponds to the most probable statistic state. The maximum can be found by the Lagrange method by setting

$$\begin{aligned}\delta H^{*\prime} &= \delta \left[-\int f_n(x) \ln f_n(x)\, dx - \alpha_0 \int f_n(x)\, dx \right.\\ &\qquad \left. - \alpha n \int \left(\frac{1}{n} \sum_{i=1}^{n} a(x_i) \right) f_n(x)\, dx \right]\\ &= -\int \delta \left[\left(\ln f_n(x) + 1 + \alpha_0 + \alpha \sum_{i=1}^{n} a(x_i) \right) f_n(x) \right] dx = 0. \end{aligned} \tag{12}$$

The parameters α_0 and α are to be determined such that (5) and (7) are satisfied. And so

$$f_n(x) = Z^{-1} \exp\left[-\alpha \sum_{i=1}^{n} a(x_i) \right], \tag{13a}$$

where
$$Z = \int \exp\left[-\alpha \sum_{i=1}^{n} a(x_i)\right] dx. \tag{13b}$$

$Z = Z(\alpha)$ is the state integral of the statistical system, and can be obtained from the equation

$$\bar{a} = \int \left(\frac{1}{n} \sum_{i=1}^{n} a(x_i)\right) Z^{-1} \exp\left[-\alpha \sum_{i=1}^{n} a(x_i)\right] dx,$$

and hence

$$\bar{a} = -\frac{1}{n}\frac{\partial}{\partial \alpha} \ln Z(\alpha) \equiv \mathfrak{H}(\alpha) \tag{14}$$

shows that the microscopic distribution depends on both α and $\bar{a}$.

If we take $a(x_i) = x_i$ with mean value $\bar{x} = n^{-1} \sum_{i=1}^{n} x_i$, then we have

$$Z = \int \exp\left[-\alpha \sum_{i=1}^{n} x_i\right] dx = \int \exp[-\alpha n \bar{x}] dx$$

$$= \left[\int_{0}^{\infty} \exp[-\alpha \bar{x}] d\bar{x}\right]^{n} = \alpha^{-n}, \tag{15a}$$

and

$$f_n(x) = Z^{-1} \exp\left[-\alpha \sum_{i=1}^{n} x_i\right] = Z^{-1} \prod_{i=1}^{n} \exp[-\alpha x_i], \tag{15b}$$

hence

$$f_n(x) = \alpha^n \prod_{i=1}^{n} \exp[-\alpha x_i]. \tag{15c}$$

Then from (15a) and (14) it follows that

$$\bar{x} = -\frac{1}{n}\frac{\partial}{\partial \alpha}[-n \ln \alpha] = \alpha^{-1}. \tag{16}$$

Using (15c) and (16), for the one-point distribution we obtain

$$f_1(x) = \bar{x}^{-1} \exp[-x/\bar{x}]. \tag{17}$$

And if we take the integrated probability function

$$F(x) = \int_{0}^{x} f_1(t) dt \quad \text{with } t \leq x,$$

then we get

$$F(x) = \int_{0}^{x} \bar{x}^{-1} \exp[-t/\bar{x}] dt = 1 - \exp[-x/\bar{x}]. \tag{18}$$

In many biological populations the distribution of the quantities is not describable by a simple exponential law. Mostly the distributions of the microscopic values are unimodal or nearly normal distributions. They may be described by the gamma distribution, which is the generalization for continuous values of the Poisson distribution. Therefore it is necessary to consider

$$a(x_i) = -\ln(x_i/\bar{x}). \tag{19}$$

The arithmetic mean of these values is

$$\bar{a} = -\frac{1}{n}\sum_{i=1}^{n} \ln(x_i/\bar{x}) = -\frac{1}{n}\ln \prod_{i=1}^{n}(x_i/\bar{x}) = -\ln\left[\prod_{i=1}^{n}(x_i/\bar{x})\right]^{1/n}, \tag{20}$$

that is, $\bar{a}$ is minus the natural logarithm of the geometric mean of $x_i/\bar{x}$. Now we proceed in the same way as above to obtain the desired result. For this, in the variation problem a new Lagrange parameter β must be introduced. The basic theorem of the variation problem shows that for $f_n(x)$ we may take the exponential function

$$f_n(x) = Z^{-1} \exp\left[-\alpha \sum_{i=1}^{n} x_i + \beta \sum_{i=1}^{n} \ln(x_i/\bar{x})\right]. \tag{21}$$

Letting $Z = z^n$, it follows that

$$f_1(x) = z^{-1}(x/\bar{x})^\beta \exp[-\alpha x],$$

$$z = \int_0^\infty (x/\bar{x})^\beta \exp[-\alpha x]\,dx. \tag{22}$$

With the help of Euler's gamma function it is possible to evaluate $z(\alpha, \beta)$:

$$z = z(\alpha, \beta) = \frac{\Gamma(\beta+1)}{\alpha(\alpha\bar{x})^\beta},$$

and hence

$$\ln z = \ln \Gamma(\beta+1) - \ln \alpha - \beta \ln \alpha\bar{x}. \tag{23}$$

The arithmetic mean and the macroparameters $a(x_i)$ can be obtained by an equation system which can be partially solved by means of Euler's psi function:

$$\bar{a} = \ln \Gamma(\beta+1) - \Psi(\beta) \equiv G(\beta), \tag{24a}$$

$$\alpha = \frac{\beta+1}{\bar{x}}. \tag{24b}$$

If $\bar{a}$ is given, β and α may be determined in a graphic way or by means of tables. Using the state integral, the one-point distribution and the probability function, the integrated probability function is found to be

$$F(x) = \frac{1}{\Gamma(\beta+1)} \int_0^{(\beta+1)(x/\bar{x})} e^{-t} t^\beta \, dt \equiv I((\beta+1)(x/\bar{x}), \beta+1). \quad (25)$$

For the solution we can use the tables given by Pagurova [1] for the incomplete gamma function

$$I(x, m) = \frac{1}{\Gamma(m)} \int_0^x e^{-t} t^{m-1} \, dt, \quad (26)$$

if we substitute $\beta + 1$ for m and $(\beta + 1)(x/\bar{x})$ for x. The probability function for various values of β has the form shown in Fig. 1.

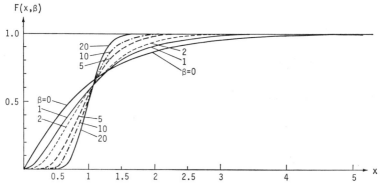

Fig. 1. The probability function $F(x, \beta)$ for various values of β.

3. The deviation parameter β

The parameter β shows the deviation of any distribution from the pure exponential distribution. When $\beta = 0$, then the function (25) reduces to the function (18). If $\beta = \infty$ ($\bar{a} = 0$), then all the elements are in the same state $x_i = \bar{x}$. Therefore β is a distinctive mark for the regulation of the state of the system. The regulation can be a gearing in some population by human influence or by other influences. For example, in medical research the distribution of any marks of the cell or the blood or other organs can be changed by the influence of some illness. Then the value of β shows the measurable intensity of the illness in comparison with the value of β for the distribution of the same marks in the healthy state or by the influence of any medicine. So the differences between the empirical distribution and the most probable distribution may be interpreted as the influence of regulation occurrences.

For other questions it is possible to reconstruct the most probable distribution of microscopic values. One can take the information which is

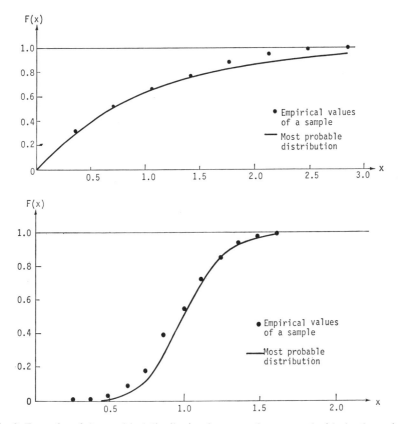

Fig. 2. Examples of the empirical distribution for a sample compared with the theoretical most probable distribution.

recessively given by the aggregated macroscopic values. If they are estimated only from a small sample, it is allowed to correct the non-significant distribution of this sample to obtain the distribution of the whole population. Until now, a very good agreement has been found between the empirical distributions of special quantities of growth taken from significant samples and the theoretically most probable distributions reconstructed on the base of the macroscopic values (see Fig. 2).

As the case may be, the method is usable for various problems, and now it is necessary to investigate their application to various biological questions.

References

[1] W. J. Pagurova, *Tables for the Incomplete Gamma Function* (Moscow, 1963) (in Russian).
[2] K. Voss, On statistical systemtheory, Dresden, 1969 (unpublished).
[3] K. Voss, On the statistical systemtheory in the economy, *Wiss. Z. Techn. Univ. Dresden* 21 (1972) 525–534.

MATHEMATICAL MODELS IN CELL CYCLE KINETICS

B. JANSSON*

Research Institute of National Defense, Stockholm, Sweden

In biology and medicine, descriptions by quantitative models are used more and more frequently. A contributory reason for this is the fact that the numerical treatment of the complex systems involved needs computer processing. Before the computer era, these types of analyses were not practicable. Cell proliferation kinetics is an example of such a field where quantitative models have been used increasingly for the last ten years.

As the cells of an organism pass from one division to the next, they go through a number of different phases of development. Since the sensitivity of the cells to radioactive radiation and to certain chemical substances varies in these phases, it is important to be able to calculate the length of the different cell phases, and to determine where they are located in the cell cycle – this applies to healthy cells as well as to malignant cancer cells. Such knowledge is important when it comes to deciding on therapeutical strategies. By introducing thymidine labelled with the radioactive hydrogen isotope tritium, Howard and Pelc [2] could show, as early as in the beginning of the 1950's, that the DNA of the cell is synthetized only in a limited stage of the cell cycle, the so-called S-phase (phase of synthesis). Another phase of the cell division cycle (the mitosis, M-phase), in which the actual division of the cell takes place and the chromosomes are visible even in a light microscope, was well-known before, and when it was found that S and M are separated in the cell cycle, there was material for a first model of the cycle (Fig. 1(a)). Howard and Pelc called the two intervals that separate M and S, and S and M, G_1 and G_2, respectively, "G" denoting "gap".

So far, the description of the cell cycle is chiefly qualitative. The important next step was to determine the length of the four consecutive phases and here also tritiated thymidine was used. After labelling cells in S-phase with tritium, the labelled cells are observed as they go through the phases of the cell cycle. In this context the most important tool for analysis is the percentage of labelled cells in mitosis (PLM = Percent Labelled Mitosis) as a function of the

* Present address: National Large Bowel Cancer Project, M. D. Anderson Hospital and Tumor Institute, Houston, Texas 77025, U.S.A.

time elapsed after labelling (Fig. 1(b)). Quastler [9] who performed the first analysis, assumed that all the cells of a cell population spent equally long times in the G_1-phase, then equally long times in the S-phase, etc., which led to a very simple deterministic model, by which the PLM curve can be drawn directly as a periodic function consisting of a sequence of congruent trapezoids. From this, the length of G_2, M, S and G_1 can be estimated, and thus also the

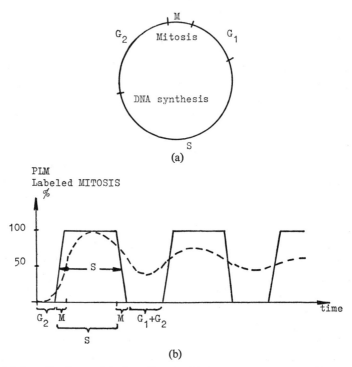

Fig. 1. (a) A qualitative model of a cell cycle. (b) PLM curves corresponding to a deterministic model (solid line) and an empirical registration (dashed line).

length of the whole cell cycle. On comparison with experimental registrations of PLM, however, the deterministic model proves to be too simple. The true curve is a damped oscillation tending towards a constant value. In most cases the curve is only in the very first oscillation reasonably in accord with its theoretical deterministic counterpart.

The weakness of Quastler's approach is obvious. The cell cycles and their various subphases vary stochastically, different cells spend different periods of time in the different phases. It has been shown in practice, however, that the thumb rules obtained by the deterministic model for estimating the mean length of the phases of the cell cycle yield fairly good estimates. One example of such an estimate is that the time between the first two 50% passages of the PLM curve can be used as a measure of the length of S.

To cope with the weakness in Quastler's model, Barrett [1] built a Monte Carlo model in which some thousands of cells and their daughter cells are observed through a number of cell generations. After having tried different statistical distributions of the time that a cell spends in one phase of the cell cycle, he found that the best result was reached when he assumed that the distribution of these transit times was log-normal. Since the middle of the 1960's, Barrett's Monte Carlo model has been in use at several cell research centres for the analysis of PLM curves. The estimation of the cell cycle parameters by comparison with experimental data requires a number of iterations with gradual improvements of the parameters. Since, due to the Monte Carlo principle, each iteration has to consist of a number of repetitions with identical parameters, the method will be time-consuming and expensive. In order to eliminate the manual treatment, an automatic estimation programme developed by Steel and Hanes [11] has been used since 1970. This programme successively estimates mean value and standard deviation first of the transit times of the G_2-phase in one region of the PLM curve, then of the S-phase in another region of the curve, and finally the mean value and the standard deviation of the transit times of the G_1-phase in a third region of the PLM registration. Then the estimation returns to G_2 in another iterative cycle, etc.

Monte Carlo models are always expensive and time-consuming, and often require a good deal of computer memory space. It is thus natural to try to replace such models by analytical models. A first attempt in that direction was made by Takahashi [12, 13], who looked upon the cell cycle as a series of subphases which the cell had to pass in proper order. A certain number of such consecutive subphases were allotted to the G_1 phase, a subsequent number of consecutive subphases to the S-phase, etc. (Fig. 2). By assuming that the time period that a cell stays in each of the subphases is exponentially distributed, it is possible to make up a system of differential equations by which the number of cells $(c_j(t), j = 1, 2, ..., N)$ in each of the subphases can be determined as a function of time. The assumption that the distribution is exponential in each subphase implies that the distribution over a whole phase (G_1, S, G_2 or M) is of gamma type and not log-normal as in the Barrett model. Results from the analytical multicompartment model agree well with experimental PLM curves, and this model has also been used successfully in estimating the parameters of the cell cycle. In order to eliminate the manual treatment, an automatic method of estimation has, also in this case, been developed and described by Takahashi, Hogg and Mendelsohn [14].

When I began my own studies in the field, I assumed that the dynamics of the cell cycle must be included in the model [3–7, 10]. A cancer cell population is nearly always in a process of change, it grows and, as a consequence, the properties of the cell cycle change; it is exposed to a treatment which reduces the number of cells, and which also changes the parameters of the cell cycle, etc. If the model is to be of use, e.g. therapeutically, the variation of the cell

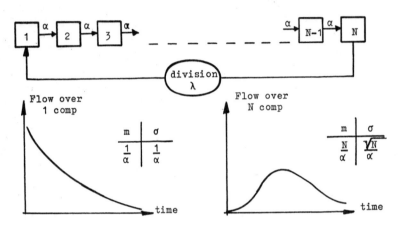

Fig. 2. Schematic representation of a multicompartmental model. (Compartments $1, \ldots, N_{G_1}$ belong to G_1, $N_{G_1} + 1, \ldots, N_S$ to S, $N_S + 1, \ldots, N_{G_2}$ to G_2, and $N_{G_2} + 1, \ldots, N$ to M.) α is the flow rate and λ the number of daughter cells to each cell at a division. The corresponding difference and differential equation systems are

$$\begin{cases} C_{1,J+1} = C_{1,J} - \alpha C_{1,J} + \lambda \alpha C_{N,J}, \\ C_{i,J+1} = C_{i,J} - \alpha C_{i,J} + \alpha C_{i-1,J}, & i = 2, 3, \ldots, N, \end{cases}$$

$$\begin{cases} \dot{C}_1 = -\alpha C_1 + \lambda \alpha C_N, \\ \dot{C}_i = -\alpha C_i + \alpha C_{i-1}, & i = 2, 3, \ldots, N. \end{cases}$$

cycle with the prevailing environment has to be taken into account. I set out from a multicompartment model analogous to that of Takahashi, which I did not know of at that time. By simulating the development of the cell population gradually in discrete time steps and using a system of difference equations, it is possible to calculate the number of cells in each subphase as a function of time.

The model was initially applied to certain types of cells of which there existed already experimental PLM analyses from various stages of growth of the tumour. The tumour studied is the ELD ascites tumour in mice, and the

data used were taken mainly from experiments by Tannock [15] and Révész and Jansson [10]. It was evident that, when the tumour grew, i.e. the number of cells grew:
  the flow rate decreased, i.e. the cells moved more slowly in the cycle,
  the number of subphases in S, G_2 and M remained unchanged,
  the number of subphases in G_1 increased.

Assuming that G_1 is the productive phase of the cell in which it performs its duties in the cell system, while S, G_2 and M together constitute its reproductive phase where the cell prepares and carries out its division, it seemed probable that the extension of G_1 would be more or less inversely proportional to the reduction of the flow rate. This proved to agree very well with experimental data, and since the flow rate as a function of the size of the tumour can be well described by a simple mathematical expression, we had all the material necessary for the formulation of an all-dynamic model. The application of the model – with an identical set of parameters – to a tumour at the age of 2, 6 and 10 days shows that the model well expresses the dynamics of the growth process (Figs. 3–5). In these figures are given both the PLM curves after a single label, and the curves of the percentage of labelled cells (PL-curves) after repeated labelling every 4th hour. The parameters were estimated from the PLM registrations and the same estimates have then been used in the comparison between theoretical and observed PL curves.

After the parameters determining the cell cycle have been estimated from PLM experiments, it is possible secondarily to calculate the total number of cells by summing up the number of cells in all the subphases. A corresponding calculation has been done experimentally by Tannock [15], and the comparison between theory and practice is shown in Fig. 6. It should be noted here that the theoretical curve is not a fitting of experimental data – it has been derived independently by means of the previous PLM analyses. Such a secondary agreement between model and reality in a sector that has not been primarily used in the construction of the model or in determining its parameters is, of course, a particularly good support for the model builder in his evaluation of the quality of the model.

The dynamic model has also allowed the description of PLM curves in an illustrative three-dimensional diagram (Fig. 7). In this the time after thymidine labelling is given, as before, on the abscissa axis, the number of cells in the tumour at the point of labelling on the ordinate axis, while the contour lines in the diagram are curves of PLM equipercentage. From this diagram it is easily seen how the cell cycle changes when the number of cells in the population increases or decreases.

The model has been used in a number of small studies. It has for instance been found that the time period during which the tritiated thymidine is active,

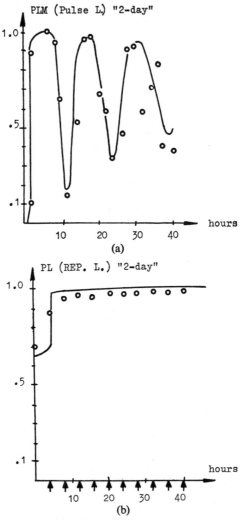

Fig. 3. Comparison between observed and theoretical PLM and PL values for a two day old tumour.

$TC = 0.08 \times 10^8$ is the total number of cells in the tumour;

$\alpha = 3.6$ is the undamped flow rate;

$\delta = 0.06 \times 10^{-7}$ is a damping constant;

$\lambda = 2$ is the number of daughter cells to each cell at a division;

$\alpha_D = 3.6(1 + 0.6 \times 10^{-7} TC)^{-1}$ is the damped flow rate determined by means of α, δ and TC;

JR = 24 min is the activity time of the tritiated thymidine;

$G_1 = 7$ comp., $S = 25$, $G_2 = 3$, $M = 1$;

α dynamic, G_1 dynamic.

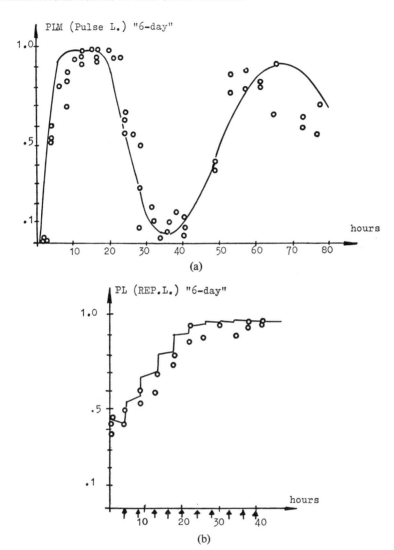

Fig. 4. As Fig. 3, though for a six day old tumour. TC = 1.4×10^8, $G_1 = 19$. The other data are as in Fig. 3.

i.e. the time interval in which it is incorporated into DNA, can be disregarded. Thus the time of activity of the tritiated thymidine, which is of the order of half an hour, can as a rule be given the value zero in theoretical models.

I shall describe in more detail an application that illustrates the importance of using a dynamic model in the study of the growth of cell populations

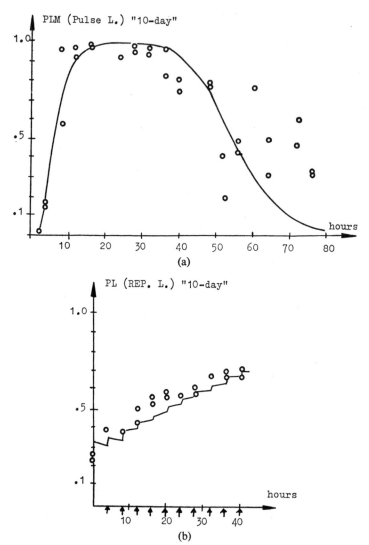

Fig. 5. As Fig. 3, though for a ten day old tumour. TC = 4.2 × 10^8, G_1 = 35. The other data are as in Fig. 3.

[3–7, 10]. When the population consists of few individuals, the growth proceeds at a maximum rate, and the flow rate of the cells in the cycle, as well as the number of subcompartments in the different cell cycle phases, are approximately constant. With constant parameters, the increase of the number of cells will be exponential, and for this ideal growth I use the mathematically flavoured connotation "growth in an infinite mouse". When the number of

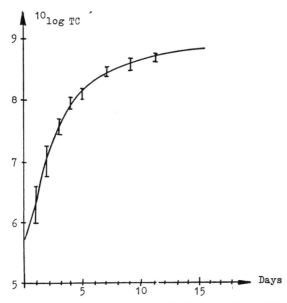

Fig. 6. Comparison between empirical and theoretical values of the total number of cells. Note that the theoretical curve is not a fitting of the experimental material, but a result of the PLM analyses. $G_{1,D} = 7(1 + 0.6 \times 10^{-7}\,TC) + 0.5$. The other data are as in Fig. 3.

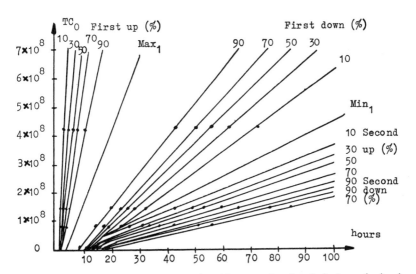

Fig. 7. Three-dimensional representation of PLM curves showing their dynamic development. Data as in Fig. 3.

cells increases, the growth will be gradually damped, and we pass from the ideal case to the real case – from the infinite mouse to the finite one, where the flow rate and number of subcompartments will change. The rate of change increases as the population increases, it reaches a maximum, and then it gradually decreases when the population grows so large that the growth practically ceases.

In the stage when the change in the parameters is great, it may involve a difference that cannot be disregarded between the values at the beginning and at the end of one and the same PLM registration. This is shown in Fig. 8, which presents two PLM curves that have the same parameters at the time of

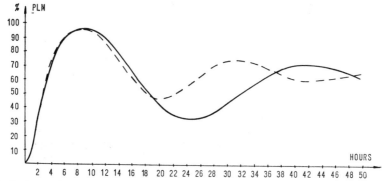

Fig. 8. Comparison between a dynamic and a static representation of a PLM curve for a medium size rapidly changing tumour; $T(0) = 2 \times 10^8$. Solid curve: dynamic; $\alpha = 1$, $\lambda = 2$, $\delta = 0.32 \times 10^{-8}$, $\beta = 0.14 \times 10^{-8}$, Dashed curve: static; $\alpha = 0.61$, $\lambda = 2$, $\delta = 0$, $\beta = 0$.

pulse labelling with tritiated thymidine. In one curve (continuous) the parameters are allowed to vary dynamically with the size of the population, whereas in the other curve (dashed) they are kept unchanged. The example is taken from a medium-sized ascites tumour with rapidly changing parameters and we see that the difference between the two curves is considerable.

Fig. 9 illustrates the effect of a dynamic approach from a different angle. Generally a tumour is heterogeneous with different conditions of growth in different parts of the tumour. Similarly, we get a heterogeneous material when transplanting a wanted number of cells from one host animal to another. The numbers of cells that really start proliferating in the new host animals will vary statistically from one animal to another. Since the growth rate depends on the number of cells, this means that we get different PLM curves for the different host animals.

If we assume that the initial values are normally distributed with a certain mean value and a certain standard deviation, the PLM curves will fall within the shaded area in Fig. 9, which is bounded by the curves that represent the mean initial value augmented and reduced by three times the standard deviation. The increase of the area of dispersion of the PLM points with increasing distance in time from the point of labelling is characteristic of PLM registrations. The figure further shows that the PLM curve corresponding to the mean initial value more or less touches the area of variation in its maximum and minimum values. A consequence of this is that if we estimate the parameters of the model by for instance a least-square method we get as a result a PLM curve, that lies too low in the maximum points and too high in the minimum points. The estimation should accordingly rather be done by a

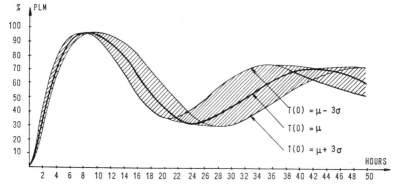

Fig. 9. Range of variation for PLM curves as a function of the statistically varying number of cells in the population at the time of labelling. T(0) is normally distributed with mean $\mu = 2 \times 10^8$ and standard deviation $\sigma = 2 \times 10^7$. $\alpha = 1$, $\lambda = 2$, $\delta = 0.32 \times 10^{-8}$, $\beta = 0.14 \times 10^{-8}$.

maximum-likelihood method that takes into account the variation of the initial values.

In the model as described here, I have assumed that the flow rate is the same over the whole cell cycle. This implies that the transit times of a cell over a given number of consecutive subphases have the same statistical distribution independently of the location of these subphases in the cell cycle. Mendelsohn and Takahashi [8] have, however, proposed the assumption that the coefficient of variation, i.e. the quotient between the standard deviation and the mean value, is the same in all four cell cycle phases G_1, S, G_2 and M.

These two assumptions can easily be compared [6]. Let

α_x = flow rate in phase x,
n_x = number of subphases in phase x,
m_x = mean transit times in phase x,
σ_x = standard deviation of the transit time in phase x, and
CV_x = coefficient of variation in phase x,

where $x = G_1, S, G_2, M$.

From the assumption of gamma distributed transit times it then follows that

$$m_x = \frac{n_x}{\alpha_x},$$

$$\sigma_x = \frac{\sqrt{n_x}}{\alpha_x},$$

$$CV_x = \frac{1}{\sqrt{n_x}}.$$

If the flow rate is constant, i.e.

$$\alpha_{G_1} = \alpha_S = \alpha_{G_2} = \alpha_M,$$

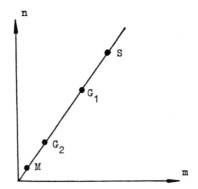

Fig. 10. If the flow rate is constant, the (m, n)-values corresponding to G_1, S, G_2 and M will be on a straight line through the origin.

then the first equation shows that the points (m_{G_1}, n_{G_1}), (m_S, n_S), (m_{G_2}, n_{G_2}) and (m_M, n_M) are on the same straight line through the origin (Fig. 10). If, on the other hand,

$$CV_{G_1} = CV_S = CV_{G_2} = CV_M,$$

it follows from the third equation that

$$n_{G_1} = n_S = n_{G_2} = n_M,$$

implying that in this case the four (m, n)-points are on a straight line parallel to the m-axis (Fig. 11).

The two assumptions have been compared by means of 80 empirical PLM analyses representing a heterogeneous cell material, in which m_x and σ_x

(or the equivalent quantities α_x and n_x) were determined independently for each phase $x = G_1, S, G_2, M$. The results of this comparison were that

the assumption of constant flow rate appears to be valid for the phases S, G_2, and M, while

the assumption of constant coefficient of variation seems to be valid for the phases G_2 and G_1.

The results can be presented in a diagram if we enter the $(\bar{m}_x, \bar{n}_x)$-points in an (m, n)-coordinate system (Fig. 12). $\bar{m}_x$ and $\bar{n}_x$ are then mean values for m_x and n_x of the 80 PLM analyses for $x = G_2, S, G_1, M$. It is evident from the figure that the points corresponding to G_2 and S are approximately on a line

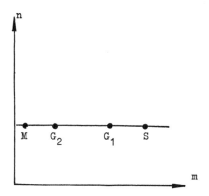

Fig. 11. If the coefficient of variation is constant, the (m, n)-values corresponding to G_1, S, G_2 and M are on a line parallel to the m-axis.

through the origin – the assumption of constant flow rate holds good – and that the points representing G_2 and G_1 are approximately on a line parallel to the m-axis – the assumption of constant coefficient of variation is acceptable.

These results lead to a modification of the original model. We may thus assume that the cell cycle should be divided into one *reproductive main phase* consisting of $G_1 + S + G_2 + M$, with the same flow rate throughout, and one *productive main phase* consisting of one single phase, which we call G_0, and which has a flow rate different from that of the reproductive phase. The original G_1-phase has thus been divided into two phases, one of which, keeping the name G_1, has been transferred to the reproductive phase, while the rest, called G_0, alone represents the productive phase.

This modified assumption has been entered into a fairly complete event-governed simulation model, which includes among other features (Fig. 13):

differentiation of cells,
competition between several types of cells,

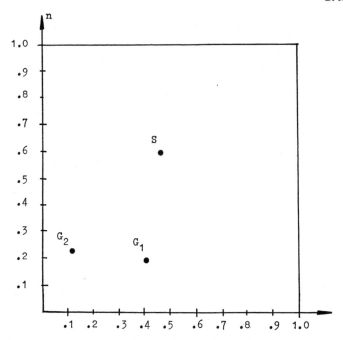

Fig. 12. The mean values of (m,n)-points for G_1, S and G_2 derived from 80 PLM analyses.

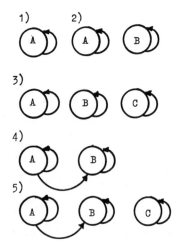

Fig. 13. Cell systems and events that can be simulated with the event governed model. A, B and C are different kinds of cells which can be studied separately or in competition with one another. They can reproduce themselves, (A), or differentiate into another type of cell, (A) (B).

labelling with one or both of two isotopes in arbitrary subphases,
killing of cells in arbitrary subphases,
transplantation of the tumour from one host animal to another.
As a fictional application of this new simulation model, I shall describe a case where we decide on a therapeutic strategy.

We assume (Table 1) that we have two kinds of cells, one consisting of malignant cells, cell type 1, and one consisting of healthy cells, cell type 2

Table 1

Basic data for exemplifying a strategy of therapy. α and β are the flow rates of the reproductive and the productive phase, respectively. TC is the total number of cells. Initial values: 10^8 cells of each type. The objective is to get rid of type 1 without permitting type 2 to be less than 10^7 cells.

	Cell type 1	Cell type 2
G_1	2	1
S	18	18
G_2	5	3
M	1	1
G_0	4	2
α	$1.22(1 + 0.7 \times 10^{-8}\,\text{TC})^{-1}$	$1.95(1 + 0.9 \times 10^{-8}\,\text{TC})^{-1}$
β	$1.22(1 + 1.4 \times 10^{-8}\,\text{TC})^{-1}$	$1.95(1 + 1.8 \times 10^{-8}\,\text{TC})^{-1}$

(continuous and dashed lines, respectively, in the figures). We also have a therapeutic agent which kills all cells in the S-phase whether they are healthy or sick. We can determine the points of time when the agent is to be used and how long it is active. The objective is to decide on a strategy of treatment such that cell type 1 disappears without cell type 2 ever falling below 10^7 cells. Before the first treatment we have 10^8 cells of each kind.

After the first treatment (dotted line in Fig. 14), the healthy cells are reduced right down to their lower limit, where the treatment has to be interrupted. In the lower part of the diagram, the percentage is given of cells in the S-phase of the two cell types as a function of the time elapsed after the interruption of the first treatment. At time = 16 we find a first occasion suitable for treatment no. 2, and at time = 45 a second one. The latter has been used in the later part of the figure with a positive but rather insignificant effect; we have managed to reduce the malignant cells (solid line) to about a third of their original number without the healthy cells decreasing in number.

If instead we use time = 16 for treatment no. 2, we obtain the result shown in Fig. 15: very effective against the malignant cells, but unfortunately

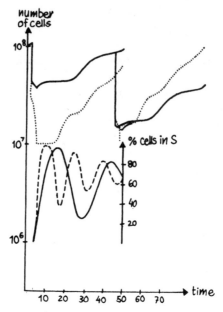

Fig. 14. For explanation see text.

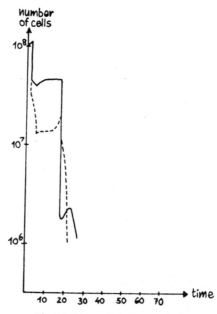

Fig. 15. For explanation see text.

with a devastating effect on the healthy cells. The poor result is obviously a consequence of too long a time of activity in our therapeutic agent. We therefore decided to shorten the time of activity as much as possible in both treatments 1 and 2. The result of this strategy is shown in Fig. 16, and, as is seen, we have now secured a very good result. From the diagram of the percentage of cells in phase S after treatment no. 2 it is seen that the next favourable occasion occurs at time = 29.

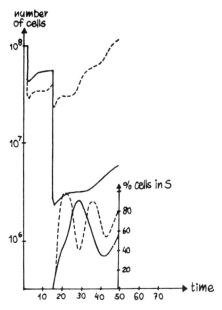

Fig. 16. For explanation see text.

Fig. 17 finally shows how this third treatment reduced the number of malignant cells to about 1/500 of their original number whereas the healthy cells never fell below 1/5 of their initial number. Since the healthy cells also decrease in number with each treatment, it is predictable that after a number of treatments the healthy cells have to be given a period of rest in order to regain their initial number. Thus the strategy will be first to use a few treatments with 13–14 hours intervals and the shortest possible time of activity, then to allow a longer period of recovery before starting a new series of treatments. This is a strategy also experimentally tried out in chemotherapy. The disastrous consequences of an error of 5–10 hours in the choice of time for treatment can also be observed in the figures. As has already been pointed out, this is a fictional example. The cell cycle data, however, have been taken from two well-known and experimentally often used types of cells – cell type 1 is the tetraploid variety, and cell type 2 the diploid type of Ehrlich's ascites tumour found in mice.

Apart from the knowledge of cell phases and their lengths there are many other factors that influence the possible effects of therapy. We have disregarded those factors in this example.

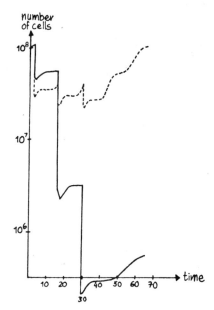

Fig. 17. For explanation see text.

References

[1] J. C. Barrett, A mathematical model of the mitotic cycle and its application to the interpretation of percentage labelled mitoses data, *J. Natl. Cancer Inst.* 37 (4) (1966) 443–450.
[2] A. Howard and S. R. Pelc, Synthesis of desoxyribonucleic acid in normal and irradiated cells and its relation to chromosome breakage, in: *Proc. Symp. on Chromosome Breakage, Heredity* 6 Suppl. (1953) 261 ff.
[3] B. Jansson, A dynamic description of the cell cycle and its phases, FOA P Rept. C8299 (November 1971).
[4] B. Jansson, A dynamic analysis of experimental PLM curves etc. for a mouse ascites tumour, FOA P Rept. C8319 (January 1972).
[5] B. Jansson, Some effects of the dynamic variation of cell cycle parameters, FOA P Rept. C8320 (February 1972).
[6] B. Jansson, Constant flow rate or constant variation coefficient – a study of some cell cycle properties, FOA P Rept. C8341 (November 1972).

[7] B. Jansson, Competition within and between cell populations, in: *The Size of Individual Dose Fractions in Radiotherapy* (Igaku-Shoin, Tokyo, 1973) 57–72.
[8] M. L. Mendelsohn and M. Takahashi, A critical evaluation of the fraction of labeled mitoses method as applied to the analysis of tumor and other cell cycles, in: R. Baserga, ed., *The Cell Cycle and Cancer* (Dekker, New York, 1971).
[9] H. Quastler, The analysis of cell population kinetics, in: L. F. Lamerton and R. J. M. Fry, eds., *Cell Proliferation* (Blackwell, Oxford, 1963).
[10] L. Révész and B. Jansson, Cell cycle analysis of mixed tumor cell populations, FOA P Rept. C8324 (1972).
[11] G. G. Steel and S. Hanes, The technique of labelled mitoses: analysis by automatic curve-fitting, *Cell Tissue Kinet.* 4 (1971) 93–105.
[12] M. Takahashi, Theoretical basis for cell cycle analysis. I. Labelled mitosis wave method, *J. Theoret. Biol.* 13 (1966) 202–211.
[13] M. Takahashi, Theoretical basis for cell cycle analysis. II. Further studies on labelled mitosis wave method, *J. Theoret. Biol.* 18 (1968) 195–209.
[14] M. Takahashi, J. D. Hogg and M. L. Mendelsohn, The automatic analysis of PLM curves, *Cell Tissue Kinet.* 4 (1971) 505–518.
[15] I. F. Tannock, A comparison of cell proliferation parameters in solid and ascites Ehrlich tumors, *Cancer Res.* 29 (1969) 1527–1534.

EVOLUTION AND CONTROL OF A BIOCHEMICAL SYSTEM

J. P. KERNEVEZ

Faculté des Sciences, Université de Dijon, Dijon, France

Introduction

The subject of this paper is the mathematical study of a biochemical system. Such systems are made in the Medical Biochemistry Laboratory of Charles Nicolle Hospital in Rouen. For the biochemical interest of these systems, see [5, 10, 11]. In [1, 2, 4, 6, 12], one can find the mathematical treatment of several similar systems from the points of view of existence and uniqueness of a solution for partial differential equations (p.d.e.) governing the system, numerical approximation of the solution, optimal control. For an introduction to nonlinear p.d.e. and optimal control of systems governed by p.d.e., we refer to [7, 8].

1. Description of the model case and equations

A membrane M separates two compartments I and II (Fig. 1). The membrane M is made of inactive protein coreticulated with an enzyme E.

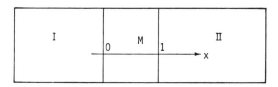

Fig. 1. A membrane M separates two compartments I and II.

At time $t = 0$, the membrane does not contain any substrate or product. Then by diffusion some substrate S comes from the compartments I and II into the membrane, where the enzyme E is a catalyst of the reaction

$$S \xrightarrow{E} P,$$

P being the product. By taking suitable units (thickness of membrane = 1, etc.), the p.d.e. governing the system is

$$\frac{\partial s}{\partial t} - \Delta s + F(s) = 0, \quad 0 < x < 1, \quad 0 < t < T,$$
$$s|_{x=0} = \alpha, \quad s|_{x=1} = \beta, \tag{1.1}$$
$$s|_{t=0} = 0,$$

where $s = s(x, t)$, $\Delta = \partial^2/\partial x^2$, α and β are positive constants, and

$$F(s) = \frac{\sigma s}{(1+|s|)}, \tag{1.2}$$

with σ a positive constant.

2. Existence, uniqueness and positivity of a solution

Let us call $s^{(0)}$ the solution of

$$\frac{\partial s^{(0)}}{\partial t} - \Delta s^{(0)} = 0,$$
$$s^{(0)}|_{x=0} = \alpha, \quad s^{(0)}|_{x=1} = \beta, \quad s^{(0)}|_{t=0} = 0, \tag{2.1}$$

and y the difference $y = s - s^{(0)}$. Then (1.1) is equivalent to

$$\frac{\partial y}{\partial t} - \Delta y + F(s^{(0)} + y) = 0,$$
$$y|_{x=0} = 0, \quad y|_{x=1} = 0, \quad y|_{t=0} = 0, \tag{2.2}$$

and if we let $\Omega = (0, 1)$, it follows from the monotonicity of F that there exists a unique solution y of (2.2) in $L^2(0, T; H_0^1(\Omega))$ (see [8]). Moreover, $s \geq 0$. This can be shown as follows: For every v in $H_0^1(\Omega)$,

$$\left\langle \frac{\partial s}{\partial t}, v \right\rangle + a(s, v) + \langle F(s), v \rangle = 0 \quad \text{a.e.} t. \tag{2.3}$$

where $\langle \ , \ \rangle$ means duality between $H^{-1}(\Omega)$ and $H_0^1(\Omega)$ or scalar product in $L^2(\Omega)$, and

$$a(s, v) = \int_0^1 \frac{\partial s}{\partial x} \frac{\partial v}{\partial x} dx. \tag{2.4}$$

We can take $v = s^-$, which gives

$$\left\langle \frac{\partial s}{\partial t}, s^- \right\rangle - a(s^-, s^-) + \langle F(s), s^- \rangle = 0. \tag{2.5}$$

Using $a(s^-, s^-) \geq 0$ and $\langle F(s), s^- \rangle \leq 0$, we get

$$\left\langle \frac{\partial s}{\partial t}, s^- \right\rangle \geq 0. \qquad (2.6)$$

Then by integration over $(0, t)$,

$$0 \leq \int_0^t \langle \partial s(\tau)/\partial t, s^-(\tau) \rangle \, d\tau = -\tfrac{1}{2} \|s^-(t)\|^2 \quad \text{in } L^2(\Omega) \qquad (2.7)$$

so that

$$s(x, t) \geq 0 \qquad (2.8)$$

a.e. in $Q = \Omega \times (0, T)$.

3. Numerical approximation of the solution

The *explicit method* is very easy to programme and is stable for

$$\frac{\Delta t}{\Delta x^2} \leq \tfrac{1}{2}(1 - \sigma \Delta t). \qquad (3.1)$$

This is the method used to solve the optimal control problem state equations in Section 4.

To obtain a numerical approximation of the solution for the model case and for other more complicated systems, the *Crank–Nicholson method* has been used:

$$s_j^{n+1} - \tfrac{1}{2} \frac{\Delta t}{\Delta x^2} (s_{j+1}^{n+1} + s_{j-1}^{n+1} - 2s_j^{n+1}) + \tfrac{1}{2} \Delta t \, F(s_j^{n+1}) = S_j^n,$$

$$S_j^n = s_j^n + \tfrac{1}{2} \frac{\Delta t}{\Delta x^2} (s_{j+1}^n + s_{j-1}^n - 2s_j^n) - \tfrac{1}{2} \Delta t \, F(s_j^n), \qquad (3.2)$$

with at every level of time the Newton linearizations:

$$s_j^{n+1,k} - \tfrac{1}{2} \frac{\Delta t}{\Delta x^2} (s_{j+1}^{n+1,k} + s_{j-1}^{n+1,k} - 2s_j^{n+1,k}) + \tfrac{1}{2} \Delta t \, (F(s_j^{n+1,k-1})$$

$$+ (s_j^{n+1,k} - s_j^{n+1,k-1}) \frac{\partial F}{\partial s}(s_j^{n+1,k-1})) = 0, \qquad (3.3)$$

$$s_j^{n+1,0} = s_j^n,$$

and at every Newton iteration overrelaxation with optimal parameter ω.

4. Optimal control

4.1. *State equations, control, constraints and cost function*

The concentration v of an activator is at our disposal in the compartments I and II (Fig. 1). This activator is diffusing in the membrane M, so that the equations governing the system are

$$\frac{\partial s}{\partial t} - \Delta s + F(s, a) = 0, \quad 0 < x < 1, \quad 0 < t < T,$$

$$F(s, a) = \sigma \frac{a}{1+a} \frac{s}{1+s},$$

$$\frac{\partial a}{\partial t} - \Delta a = 0,$$

$$s|_{x=0} = \alpha, \qquad s|_{x=1} = \beta, \tag{4.1.1}$$
$$a|_{x=0} = v_0(t), \qquad a|_{x=1} = v_1(t),$$
$$s|_{t=0} = 0,$$
$$a|_{t=0} = 0.$$

If we let $\mathcal{U} = L^2(0, T) \times L^2(0, T)$, and $v = (v_0, v_1)$, then v is the control and it is constrained to be in $\mathcal{U}_{ad}$, where

$$\mathcal{U}_{ad} = \{v = (v_0, v_1) : v \in \mathcal{U}, \ 0 \leq v_i \leq M, \ i = 0, 1\}. \tag{4.1.2}$$

We want to regulate the concentration of substrate at $x = \frac{1}{2}$, that is, to minimize the cost function

$$J(v) = \int_0^T |s(\tfrac{1}{2}, t) - z_d(t)|^2 \, dt \tag{4.1.3}$$

when v spans $\mathcal{U}_{ad}$.

4.2. *Existence of at least one optimal control*

a is defined by transposition

$$\int_Q a(-\partial q/\partial t - \Delta q) \, dx \, dt = -\int_\Sigma v \, (\partial q/\partial v) \, d\Sigma \tag{4.2.1}$$

for all $q \in Z$, where $a \in L^2(Q)$ and

$$Q = \Omega \times (0, T), \qquad \Sigma = \{0, 1\} \times (0, T),$$

$$\frac{\partial}{\partial v} = \text{normal derivative directed outwards } \Omega, \tag{4.2.2}$$

$$\int_\Sigma v \frac{\partial q}{\partial v} \, d\Sigma = \int_0^T \left(v_0(t) \left(\frac{-\partial q(0, t)}{\partial x} \right) + v_1(t) \frac{\partial q(1, t)}{\partial x} \right) dt,$$

and
$$Z = \{q: q, \partial q/\partial x, \partial^2 q/\partial x^2, \partial q/\partial t \in L^2(Q), q|_\Sigma = 0, q|_{t=T} = 0\}. \quad (4.2.3)$$

Let (v^n) be a minimizing sequence:
$$J(v^n) \to \inf_{v \in \mathcal{U}_{ad}} J(v). \quad (4.2.4)$$

There exists a subsequence, always called (v^n), such that:
  (i) $v^n \rightharpoonup v$ in $\mathcal{U}$, weakly ($v \in \mathcal{U}_{ad}$);
  (ii) $a^n \rightharpoonup a$ in $L^2(Q)$, weakly;
  (iii) $a^n \to a$ in $L^2(Q)$, strongly [9];
  (iv) $s^n - s \to 0$ in $L^2(0, T; H_0^1(\Omega))$, strongly;
  (v) $s^n(\tfrac{1}{2}, \cdot) \to s(\tfrac{1}{2}, \cdot)$ in $L^2(0, T)$, strongly;
  (vi) $J(v^n) \to J(v)$;
here a^n, s^n and a, s are related to v^n, v as in (4.1.1).

There is no apparent reason for having a unique optimal control.

4.3. *Use of a Lagrangian to get the gradient of the cost function*

We define the Lagrangian

$$\mathcal{L}(v, s, a; p, q) = \tfrac{1}{2} \int_0^T (s(\tfrac{1}{2}, t) - z_d(t))^2 \, dt + \int_Q p\left(\frac{\partial s}{\partial t} - \Delta s + F(s, a)\right) dx \, dt$$
$$+ \int_Q a\left(-\frac{\partial q}{\partial t} - \Delta q\right) dx \, dt + \int_\Sigma v \frac{\partial q}{\partial \nu} \, d\Sigma \quad (4.3.1)$$

for every $v \in \mathcal{U}$, $s \in s^{(0)} + Y$, $a \in L^2(Q)$, $p \in L^2(Q)$, $q \in Z$, where

$$Y = \{y: y, \partial y/\partial x, \partial^2 y/\partial x^2, \partial y/\partial t \in L^2(Q), y|_\Sigma = 0, y|_{t=0} = 0\}. \quad (4.3.2)$$

Of course
$$J(v) = \mathcal{L}(v, s(v), a(v); p, q), \quad (4.3.3)$$

$s(v)$ and $a(v)$ being defined by (4.1.1).

It can be proved [1] that the following applications are Fréchet differentiable:

$$a \to \int_Q p F(s, a) \, dx \, dt \quad \text{from } G \text{ to } \mathbf{R}, \quad (4.3.4)$$

$$s \to \int_Q p F(s, a) \, dx \, dt \quad \text{from } G \cap (s^{(0)} + Y) \text{ to } \mathbf{R}, \quad (4.3.5)$$

$$a \to s \text{ from } G \text{ to } s^{(0)} + Y, \quad (4.3.6)$$

where
$$a \in G \Leftrightarrow a \in L^2(Q) \text{ and there exists } \delta > 0 \text{ such that } a > -\tfrac{1}{2} + \delta, \delta \in \mathbf{R}. \tag{4.3.7}$$

This gives sense to
$$\langle J'(v), \delta v \rangle = \left\langle \frac{\partial \mathscr{L}}{\partial v} + \left(\frac{\partial \mathscr{L}}{\partial s} \cdot \frac{\partial s}{\partial a} + \frac{\partial \mathscr{L}}{\partial a} \right) \cdot \frac{\partial a}{\partial v}, \delta v \right\rangle \tag{4.3.8}$$

for every δv in $\mathscr{U} = L^2(\Sigma)$.

We have
$$\langle J'(v), \delta v \rangle = \left\langle \frac{\partial \mathscr{L}}{\partial v}, \delta v \right\rangle = \int_\Sigma \frac{\partial q}{\partial v} \delta v \, d\Sigma, \tag{4.3.9}$$

$$J'(v) = \frac{\partial q}{\partial v} \quad \text{on } L^2(\Sigma), \tag{4.3.10}$$

if p and q are chosen such that
$$\frac{\partial \mathscr{L}}{\partial s} = 0, \quad \frac{\partial \mathscr{L}}{\partial a} = 0, \tag{4.3.11}$$

i.e.,
$$\int_0^T (s(\tfrac{1}{2}, t) - z_d(t)) \varphi(\tfrac{1}{2}, t) \, dt + \int_Q p \left(\frac{\partial \varphi}{\partial t} - \Delta \varphi + \frac{\partial p}{\partial s} \varphi \right) dx \, dt = 0, \tag{4.3.12}$$

$$\int_Q p \frac{\partial F}{\partial a} \psi \, dx \, dt + \int_Q \psi (-\partial q/\partial t - \Delta q) \, dx \, dt = 0 \tag{4.3.13}$$

for all $\psi \in L^2(Q)$. (4.3.12) and (4.3.13) imply
$$-\frac{\partial p}{\partial t} - \Delta p + \frac{\partial F}{\partial s} p = -(s(\tfrac{1}{2}, \cdot) - z_d(\cdot)) \otimes \delta_{x=1/2},$$
$$p|_\Sigma = 0, \quad p|_{t=T} = 0, \tag{4.3.14}$$

$$q \in Z,$$
$$-\frac{\partial q}{\partial t} - \Delta q + \frac{\partial F}{\partial a} p = 0. \tag{4.3.15}$$

Theorem 4.3.1. *The gradient $J'(v)$ of the cost function J is given by (4.3.10) if q is defined by the state equations (4.1.1) and the co-state equations (4.3.14) and (4.3.15).*

4.4. A necessary condition of optimality

If $u(=(u_0, u_1))$ is an optimal control, then
$$\langle J'(u), v - u \rangle \geq 0 \tag{4.4.1}$$

for all $v \in \mathcal{U}_{ad}$, which is equivalent to

$$u = P_{\mathcal{U}_{ad}}(u - \rho J'(u)) \quad (\rho > 0), \tag{4.4.2}$$

or

$$\begin{aligned} u_0(t) &= P_{[0,M]}\left(u_0(t) + \rho \frac{\partial q(0,t)}{\partial x}\right), \\ u_1(t) &= P_{[0,M]}\left(u_1(t) - \rho \frac{\partial q(1,t)}{\partial x}\right), \end{aligned} \tag{4.4.3}$$

$P_{\mathcal{U}_{ad}}$ being the projection from $\mathcal{U}$ on $\mathcal{U}_{ad}$, $P_{[0,M]}$ the projection from $\mathbf{R}$ on the interval $[0, M]$.

4.5. Numerical algorithm of optimization

This is the simple gradient algorithm:
Given $v^{(0)}$ (for instance $v_0^{(0)} = v_1^{(0)} = M$) and $v^{(n)}$, define $v^{(n+1)}$ by

$$v^{(n+1)} = P_{\mathcal{U}_{ad}}(v^{(n)} - \rho J'(v^{(n)})), \quad \rho > 0, \tag{4.5.1}$$

i.e.

$$\begin{aligned} v_0^{(n+1)}(t) &= P_{[0,M]}\left(v_0^{(n)}(t) + \rho \frac{\partial q^{(n)}(0,t)}{\partial x}\right), \\ v_1^{(n+1)}(t) &= P_{[0,M]}\left(v_1^{(n)}(t) - \rho \frac{\partial q^{(n)}(1,t)}{\partial x}\right), \end{aligned}$$

where $q^{(n)}$ is defined from $v^{(n)}$ by the state equations (4.1.1) and the co-state equations (4.3.14) and (4.3.15). Of course the numerical values of $v^{(n)}$ are computed at discrete times $k \Delta t$ only ($k = 0, 1, 2 \ldots$) and the state and co-state equations are solved using for instance the explicit method (Section 3).

When ρ is small, this algorithm converges slowly. When ρ is too large, it is unstable. So the main difficulty is to find a good value of ρ. This can be done at every iteration of the algorithm by polynomial interpolation. With such a change of $\rho(=\rho^{(n)})$, the convergence of the algorithm is fast. Moreover, with a "reasonable" interval $[0, M]$ for the values of the control, the limit u of $v^{(n)}$ does not depend upon the initial value $v^{(0)}$, and it seems that the optimal control is locally unique.

5. Conclusion

We have seen on a simple example that it is possible not only to compute the evolution of the profiles of concentration inside an artificial membrane, but also to look for an optimal control of a concentration inside the membrane. The method of optimization used here is quite general and has also been used in problems of identification of parameters [3].

References

[1] J. C. Brauner and P. Penel, Sur le contrôle optimal de systèmes non linéaires de biomathématique, Thesis, University of Paris (1972).
[2] D. Dubus, Evolution d'un système biomathématique, Thesis, University of Paris (1972).
[3] G. Joly, Identification de paramètres cinétiques dans un système biochimique régi par des équations aux dérivées partielles, Thesis, University of Paris (1974).
[4] J. P. Kernevez, Evolution et contrôle de systèmes biomathématiques, Thesis, University of Paris (1972); CNRS No. A.O.7246.
[5] J. P. Kernevez and D. Thomas, Diffusion reaction in enzymatically active model-membranes, Journées d'Informatique Médicale, IRIA (1972).
[6] J. P. Kernevez and D. Thomas, *Analyse et contrôle de systèmes biochimiques* (Dunod, Paris, 1974).
[7] J. L. Lions, *Optimal Control of Systems Governed by Partial Differential Equations* (Dunod, Paris, 1968).
[8] J. L. Lions, *Quelques méthodes de résolution des problèmes aux limites non linéaires* (Dunod, Paris, 1969).
[9] J. L. Lions, Optimal control of distributed parameter systems, Summer Conf., University of Maryland (1971).
[10] D. Thomas, Etude de modèles biologiques structurés à l'aide de membranes porteuses d'enzymes réticulées, Thesis, University of Rouen (1971).
[11] D. Thomas and J. P. Kernevez, Immobilized enzymes, in: *Proc. Engineering Foundation Conf.*, August 1971 (Henniker, 1971).
[12] J. P. Yvon, Etude de quelques problèmes de contrôle pour des systèmes distribués, Thesis, University of Paris (1972).

DISCUSSION

Reich:

Can the method be worked out to yield also statistical estimates of the parameters to be controlled, i.e. their range of variation rather than only the "optimal" value?

Kernevez:

No. The parameters to be controlled have definite values for which we compute approximations as close as possible, and we take a test, to stop the computations, of the forms

$$J(v^{(n)}) < \varepsilon \quad \text{or} \quad \left|\frac{J(v^{(n+1)}) - J(v^{(n)})}{J(v^{(n)})}\right| < \varepsilon \quad \text{or} \quad n < N_{\max}.$$

But the method can be used to find optimal control when the system is stochastic. What is minimized then is

$$J(v) = \sum_{i=1}^{\tau} p_i j(v;i).$$

where $j(v;i)$ is the cost function for control v and for process number i (with probability p_i, $\sum_{i=1}^{\tau} p_i = 1$).

REFLECTIONS ON SOME ALGORITHMS FOR THE SOLUTION OF THE INVERSE PROBLEM (IDENTIFICATION OR ADJUSTMENT) FOR LINEAR COMPARTMENTAL MODELS

C. MONOT*

Centre National de la Recherche Scientifique, Nancy, France

and

J. MARTIN**

Section d'Informatique Médicale, Faculté de Médecine, Université de Nancy, Nancy, France

1. Introduction

The simulation of biological phenomena has considerably developed during recent years. Mathematical models, indeed, enable testing the validity of hypotheses on the relations between the phenomena, the influence of various parameters on the observed variations, and estimation of the values of data not directly accessible to experiments, etc.

From the mathematical point of view, all this means calculating various curves from the numerical coefficients of the model, which can afterwards be compared with the true experimental results. This is the direct problem of simulation.

Experience has shown to us that this is a useful utilization, although secondary in comparison with the major interest of the "inverse problem", that is, estimation of the parameters of the model selected giving the best assessment of the experimental curves. Indeed, when a model correctly represents biological phenomena [5, 9, 10, 12, 13, 16], its parameters have a significant value which is much higher than morphological accidents of curves, which are all that have been studied up till now. The mathematical

---

* Research worker, C.N.R.S.
** INSERM Research Group U.115.

problem is more difficult. We present here various methods for its solution, attempting to show the difficulties which we encountered.

We shall limit the study to the case of models shown by a system of first-order differential equations with constant coefficients. They are greatly utilized, since they are connected with the fundamental hypothesis that exchanges of a product are proportionate to its quantity in the original compartment [17].

Example: simplified model of calcium.

Prescribed entries or exits exterior to the system:

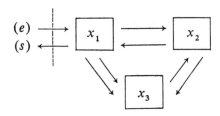

$$\frac{dx_1}{dt} = -(k_{12} + k_{13})x_1 + k_{21}x_2 + k_{31}x_3 + (e(t) - s(t)).$$

The general formulation results in the differential equation system

$$\frac{d}{dt}[X] - [A][X] = [E/S], \qquad (1)$$

where

[A] = exchange matrix, which is dependent on the exchange coefficients k_{ij},

[X] = vector of quantities contained in the various compartments,

[E/S] = vector of external prescribed movements (dependent on time, independent of the values of x_1, x_2, x_3). We shall not consider E/S in the following report. In fact, entries generally correspond to the initial conditions of the system or to a second simple member. Prescribed exits are very rare.

2. The inverse problem

2.1. *Definition*

The inverse problem is defined as research for the best numerical values of the k_{ij} parameters from experimental knowledge of one or more functions of time $x_i(t)$.

2.2. Note concerning the nature of solutions of differential equation systems with constant coefficients

Any solution of the differential equation system (1) with right-hand side put to zero is of the form

$$[X] = \sum_i [V_i] \exp \lambda_{it}, \qquad (2)$$

in which λ_i and $[V_i]$ are respectively the eigenvalues and eigenvectors of the matrix A. (In biology the real part of eigenvalues is always negative.)

Notation. In the following we shall call the elements of V_i exponential *coefficients* and the λ_i exponential *exponents*.

This formulation explains the interest taken by biologists and physicians in the sums of exponentials, whose mathematical handling is known to be difficult since they do not form a base [18].

In the following we shall call the whole of the elements of $[X]$ a *solution of the system*. A particular curve X_j will therefore be an "element" of the solution. These elements of the solution are not independent since in all compartments matter is preserved whatever the exchanges, so that at any time we have

$$\sum_i X_i(t) = \sum (\text{entries}) - \sum (\text{exits}) + \sum_i X_i(0). \qquad (3)$$

2.3. Data available for the solution of the inverse problem

If we have all the elements of information, we should know the curves of $n - 1$ compartments (because of (3)) in the case of a system with n compartments. Unfortunately the biological reality is quite different and much less favourable.

2.3.1. Knowledge of the curves as a whole

Generally, it is impossible to take measurements in all compartments which are theoretically envisaged. In a large number of cases satisfactory measurements exist for only one or two compartments: blood compartment, and sometimes urinary compartment. Reference should be made, however, to the possibility of obtaining isolated measurements (for example, for a biopsy at a certain moment) for some of the other compartments. We shall see that it is sometimes difficult to incorporate these single measurements in certain methods of solution of the inverse problem.

2.3.2. Precision of the curves

The experimental "curve", moreover, is often incomplete because of strong biological requirements.

There is sometimes continuous recording during the initial period of the phenomenon (half an hour to one hour generally). After this, there are only a few points of increasingly unequal distribution.

As an example, in a complete thyroid exploration in man, there is continuous recording for the first half-hour, and measurements after 1 hour, 2 hours, 6 hours, 1 day, 3 days, 8 days. However, certain methods suppose that the compartment of curves is known ad infinitum! Obviously it is theoretically certain that the curve is asymptotic to the time axis (or to a parallel line) for very large t, but it is impossible to know the law of approximations.

This explains the difficulty of finding a "good algorithm" of calculation. The most powerful mathematical methods are no substitute for a lack of data. Unfortunately, some people imagine the contrary to be true. We should not forget that the value of each point is tainted with error, sometimes connected with the type of measurement itself, as in the case of radioactivity.

The following problem, which to our knowledge has not been solved, is interesting: taking into account the structure of the model, what is the influence of the facts on the exact knowledge of the parameters? It would also be useful to be able to optimize the distribution of the points of measurement that can reasonably be found in order to obtain the best information as to parameters.

3. Some solution methods

We need not only a method which is theoretically satisfactory (and from this point of view the inverse problem is fairly standardized), but an efficient algorithm which converges rapidly and which is stable whatever the initial values of iteration at the outset, and capable of being put to work on a not-too-large computer. In other words, it must be possible to utilize in routine practice the numerous functional explorations (thyroid gland, kidneys, heart, iron metabolism, calcium metabolism, etc.).

In the following we shall refer only to the methods with which we have experimented, leaving aside other methods such as those based on random numbers (Monte Carlo method):

(i) calculation of the k_{ij} after approximation of the experimental function curve by a linear combination of exponentials;

(ii) various methods of optimization in which identification of the system is carried out without preliminary approach of the experimental curves;

(iii) the method using the Laplace transformation (equation) which, according to the case, utilizes approximation or otherwise, but not necessarily by means of a sum of exponentials.

3.1. Direct calculation after approximation of the experimental function by a linear combination of exponentials

3.1.1. Principle

The choice of the approximation function is imposed by the nature of the solution of the differential equation systems to which we have referred (eq. (2)).

The knowledge of a single experimental curve X_c should permit the eigenvalues and the elements corresponding to the eigenvectors to be determined. In fact, it is sufficient to adjust this curve (or its experimental points) by a sum of n exponentials (where n is the number of compartments of the system), and thus to determine the $2n$ values (n coefficients and n exponents) in a general manner. That is, to assume that the curve is known by at least $2n$ exact points.

We shall call the coefficients a_i and the exponents of exponentials b_i, as estimated from the calculation. The identification from X_c of the obtained numerical values therefore leads to two groups of equations:

(i) n relations deduced from the eigenvalues:

$$\lambda_i(k_{ij}) = b_i;$$

(ii) n relations deduced from the element c of each of the eigenvectors:

$$(V_i(k_{ij}))_c = a_i.$$

Remark 1. Among the latter, $n - 1$ relations at the maximum are independent. There is, indeed, a constraint imposed by the problem as an initial condition:

$$X_c(0) = \sum_{i=1}^{n} a_i.$$

Remark 2. The knowledge of a second experimental function X_d only affords $n - 1$ independent relations in the second group. In fact, theoretically the eigenvalues have been deduced from the first curve. However, taking into account the incertitude as to the points of measurement, the λ_i are more accurately estimated when the number of curves is higher.

3.1.2. *Maximum number of calculable parameters*

The maximum number of k_{ij} parameters which can be calculated from independent experimental m curves is therefore equal to

$$N_{max} = n + m(n-1).$$

However, in each problem other relations imposed by the structure of the model may interfere and decrease this number: eigenvalue zero, etc. This is a general restriction, independent of the method utilized.

3.1.3. *Practical use of the method*

For the true calculation, other standard relations are utilized which are easier to handle; in particular, the relations between the roots of a polynomial.

3.1.3.1. *Relations between the roots of a polynomial.* We identify the successive terms of the development of the characteristic polynomial composed for the calculation of the eigenvalues, of the matrix A,

$$|\lambda[I] - [A]| = 0,$$

and of the polynomial connecting the known roots b_i.

$$(\lambda - b_1)(\lambda - b_2)\ldots(\lambda - b_n) = 0.$$

We then obtain the relations

$$\sum_{i=1}^{n} A(i,i) = \sum_{i=1}^{n} b_i,$$

$$\sum_{\substack{i,j=1 \\ i \neq j}}^{n} (A_{ii}A_{jj} - A_{ij}A_{ji}) = \sum_{\substack{i,j=1 \\ i \neq j}}^{n} b_i b_j,$$

$$\vdots$$

$$D(A) = b_1 b_2 \ldots b_n.$$

Let us recall also that by the structure of A,

$$A_{ii} = -\sum_{\substack{j=1 \\ i \neq j}}^{n} k_{ij} \quad \text{(exits from compartment } i\text{)},$$

$$A_{ij} = k_{ji} \quad \text{(transfer } j \to i\text{)}.$$

3.1.3.2. The second group of equations deduced from the eigenvectors may be replaced by equations obtained by composing the initial conditions fulfilled by the $n - 1$ first derivatives of each experimentally known function, that is,

$$\frac{dX_c(k_{ij})}{dt} = \sum_{i=1}^{n} a_i b_i,$$

$$\frac{d^2 X_c(k_{ij})}{dt^2} = \sum_{i=1}^{n} a_i b_i^2,$$

$$\vdots$$

and, when $t = 0$

$$\frac{\mathrm{d}^{n-1} X_c(k_{ij})}{\mathrm{d}t^{n-1}} = \sum_{i=1}^{n} a_i b_i^{n-1}.$$

With the exception of a few simple cases (small number of compartments, etc.) these two groups of equations as a whole form a non-linear system. Standard methods of numerical analysis must therefore be resorted to for the solution (Newton's method, for example) [14].

3.1.4. Criticism of the method

Difficulties are essentially due to approaching the experimental function by a linear combination of exponentials.

When the number of exponentials is too great and when the exponents are too closely related, the approximations hardly converge. It should be remarked that for arbitrary initial values it is much easier to improve successively each exponent of the exponentials taken separately than to seek simultaneously for all the coefficients a_i and b_i. However, in certain cases we can obtain several groups of values of a_i and b_i so that the functions $\sum a_i \exp(b_i t)$ both approach the given function (see Appendix A.1). It should not be forgotten that although in theory the points are assumed to be known exactly and the number of exponentials well defined, in practice the measurements are uncertain (we shall refer to this later) and hypotheses must be made as to the structure of the model [5, 10].

Under these conditions it becomes difficult to isolate the absolute minimum of the sum of the differences between the local minima. Even in cases where the solution is mathematically shown to be unique [1, 2, 18], in practice we are likely to obtain several groups of parameters as possible solutions, at least with precision comparable to that obtained in practice (Appendix A.2).

However, this method is very widely used in routine for simple models. Splitting the experimental function into 1 or 2 exponentials is almost always done by hand on semi-logarithm paper or by computer, using similar methods [19]. This gives useful results, moreover, particularly when the exponents of the exponentials are sufficiently different and the number of exponentials is three at most. The resolution of the equations then leads to explicit formulae for the parameters when the system is soluble by mathematics. It should be pointed out that, in certain cases, statistical correlations are sufficiently significant in pathology.

Remark 1. We have drawn attention to the incertitude of the order of the model. Methods have been developed in this field enabling this characteristic to be identified. Unfortunately these results, although excellent in theory, assume that a large number of points of measurement are known with great exactitude [3]. This is possible in physics, but not in biology or in medicine.

The same holds true for methods developed for knowing the number of exponentials of a curve, the sum of the exponentials from exactly known points [14, 15] and from points known with a Poisson-type error [2].

Remark 2. We have seen that the number of calculable parameters is limited to a maximum value. Supplementary information may sometimes increase this number, as shown by the example of a simplified model of the kinetics of labelled iron in man:

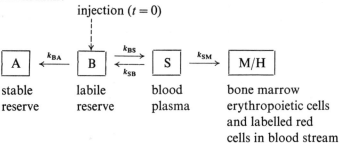

Labelled iron, injected in the initial time in the plasma, spreads into the reserves and the erythropoietic complete (M/H). Experimentally we know the plasma decrease:

$$S = a_1 \exp(b_1 t) + a_2 \exp(b_2 t), \qquad a_1 + a_2 = S(0).$$

We can determine only a maximum of three parameters, and the model comprises 4.

Experimentally the value of the level of the curve of globular incorporation (which may here be considered to be $(M/H)(\infty)$) is known:

$$M(\infty) = \int_0^\infty k_{SM} S(t)\, dt = -k_{SM}\left[\frac{a_1}{b_1} + \frac{a_2}{b_2}\right] = F.$$

Calculation of the parameters then leads to

$$k_{SM} = -\frac{F b_1 b_2}{a_1 b_2 + a_2 b_1}, \qquad k_{SB} = -k_{SM} - (a_1 b_1 + a_2 b_2),$$

$$k_{BS} = \frac{(a_1 b_1 + a_2 b_2)(a_1 b_2 + a_2 b_1) - b_1 b_2}{k_{SB}},$$

$$k_{AB} = -(a_1 b_2 + a_2 b_1) - k_{BS}.$$

Remark 3. Several different models may be satisfactory when all the independent curves are not known. Thus the mamillary model below, with the same given information $S(t)$ and $M(\infty)$ is equivalent to the preceding for the curve with which we are concerned, that is, $M(t)$:

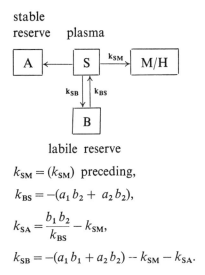

$k_{SM} = (k_{SM})$ preceding,

$k_{BS} = -(a_1 b_2 + a_2 b_2)$,

$k_{SA} = \dfrac{b_1 b_2}{k_{BS}} - k_{SM}$,

$k_{SB} = -(a_1 b_1 + a_2 b_2) - k_{SM} - k_{SA}$.

For a given curve S, in the absence of information as to A and B, a group of values of the different parameters is obtained for each system. In order to decide in favour of one or another model, data as to A or B must be known.

Remark 4. The precision of the measurements also depends sometimes on conditions over which there is no control. The best example is that of radioactivity, so widely spread in the studies of compartmental kinetics, since each measurement is a Poisson-type achievement. In the calculations, only an estimate of the real value and of its standard deviation can be used. The criteria of approximation must take this into account. For this purpose, we have perfected methods utilizing the "maximum likelihood". When there is only one exponential, this method, although somewhat awkward, gives excellent results. When there are two exponentials, it becomes extremely difficult. We have been unable to make it converge when there are more than two exponentials [2]. In practice this is serious, since simulation shows that the results obtained by the maximum likelihood method are much closer to true results than those obtained by the usual method using the criterion of least squares on the values or the logarithm of the values. This raises the entire problem of the use of a criterion of approximation, a subject which we do not wish to broach here. In practice we naturally resort to the method of least squares, but it is important to define its limitations.

3.2. The methods of optimization

3.2.1. Principle

The expression "optimization" includes a number of mathematical methods by which, from more or less arbitrary data, an attempt is made to get closer to the solution at each reiteration. An approach criterion is therefore necessary. An attempt is generally made to minimize all the variations between the experimental data and the corresponding values calculated by the differential equation system, acting on the sought parameters k_{ij}.

The "cost function" is often taken as a quadratic form

$$\Omega(k_{ij}) = \sum_{i=1}^{p} \alpha_i (f_i(k_{ij}) - F_i)^2,$$

where

$f_i(k_{ij})$ = value calculated as a function of the experimentally known parameters of the ith magnitude,

F_i = experimental numerical datum of this magnitude,

α_i = balancing factor, arbitrary or otherwise, sometimes necessary to establish a balance in the importance of data.

It should be remarked that resolution of the differential system is necessary for each calculation of this variation function Ω. The standard Kutta–Runge method proves to be rapid and effective for this purpose.

3.2.2. Principal methods

We shall mention the following methods without dwelling on their theoretical side.

3.2.2.1. *The standard method of derivative from the criterion of the "least squares"*. The values of the parameters k_{ij} are looked for in such a way that the partial derivatives of the quadratic form are zero,

$$\left[\frac{\partial \Omega}{\partial k_{ij}}\right] = 0.$$

It is then sufficient to resolve this system, which is generally nonlinear, by a standard method (e.g. Newton method).

3.2.2.2. *The gradient method.* From a point $M^{(0)}$ corresponding to arbitrary initial values of the parameters k_{ij}, the minimum of the function on the direction of the gradient in $M^{(0)}$ is looked for, that is $M^{(1)}$ corresponding to $k_{ij}^{(1)}$, and this point being found, the process is recommenced, etc.

3.2.2.3. *The method of orthogonal directions.* Blocking all parameters except one, the latter is given a value which makes Ω minimum. Permutation enables each parameter to be improved successively. The method of *local variations* is based on the following principle: to modify a parameter, say k_{ij} for example, 3 values are attributed to it successively: k_{ij}, $k_{ij} + u$ and $k_{ij} - u$, where u is an arbitrary datum. The value retained will be that which makes Ω minimum [1].

There is also the method of conjugated directions, as well as other methods, which are all derived from the preceding.

General remark. These methods can indeed be used for obtaining the approximation function $\sum_i a_i \exp(b_i t)$ of the preceding section. There again it is interesting to improve the parameters k_{ij} by small successive groups.

3.2.3. Critical analysis

These methods require the calculation of Ω to be carried out many times and thus the resolution of the differential system at each stage. The calculation time is quite long. Flexibility of use is considerable, since it is very easy to introduce a singular experimental fact: a unique point on a curve, a combination of several magnitudes, etc. Despite the attractive appearance of these calculation techniques, they often require initial values which are not too far removed from the solution. Once the system exceeds 4, convergence is often difficult.

3.3. Methods utilizing the Laplace transformation

3.3.1. Principle

The Laplace transformed curve of X_i is given by the equation

$$L_p(x_i) = \int_0^\infty e^{-pt} X_i(t)\, dt,$$

in which p is the parameter of the transformation. We shall see its importance later.

The initial differential equation system (1), after the Laplace transformation, becomes the linear system

$$[p[I] - [A]][L_p(X)] = [X(0)], \qquad (4)$$

where
  $[X(0)]$ = vector X at time $t = 0$ (the initial conditions of the problem must therefore be known),
  $[I]$ = unit matrix.

3.3.2. *Procedure for calculation*

Let us say $[B] = [p[I] - [A]]^{-1}$. The linear system (4) can then be written as

$$[L_p(X)] = [B][X(0)]. \tag{5}$$

Taking into consideration the correspondence between a function and its transformed curve, it is possible to calculate the parameters k_{ij} from eq. (5). For this we must assume that the values of the transformed curves of the experimental curves are known [11], and therefore that there is a preliminary approach of the known curves by experimental points. Furthermore, the participation of the various parts of the curves in the value of the transformed curve greatly depends on the values of p, and it is possible to use this property to increase the importance of the known parts of the curves [9, 11]. Two cases then occur.

In the first, the curves are known for $n - 1$ compartments, and only $n - 1$ coefficients k_{ij} are to be determined. In this case, if we call the curve of the ith compartment $Y_i(t)$, and its transformed curve $L_p(Y_i)$, we obtain the system

$$[L_p(X_i(k_{ij}))] = [L_p(Y_i)]. \tag{6}$$

We shall not insist on the solution of this system, because although it is standard, we have never come across it.

The second case, much more important in practice, is that in which we have only a few curves, and, at the limit, only one. In this case [11], we resolved the problem by taking as many different values of the parameter p of the transformed curve as the number (say m) of coefficients k_{ij} sought for. Let us identify these values with the elements deduced from eq. (5). We obtain an equation system

$$\begin{aligned}[L_{p1} X_1(k_{ij})] &= [L_{p1}(Y)], \\ [L_{p2} X_1(k_{ij})] &= [L_{p2}(Y)], \\ &\vdots \\ [L_{pm} X_1(k_{ij})] &= [L_{pn}(Y)]. \end{aligned} \tag{7}$$

The solution of this system of implicit equations leads to the values of the coefficients k_{ij}. The programming is somewhat complex since it requires successive working on m systems of type (5) to obtain eq. (7).

A variant of the preceding procedure consists in taking more values of p than unknown quantities k_{ij}. This comes within the scope of superdimensioned systems [4].

Remark 1. This method of calculation easily enables a second member (E/S) of the differential system to be taken into consideration if it is constant or a function of time only. For example, in medicine, intravenous drip corresponds to an entry of a known quantity of substance. It then suffices to replace eq. (5) by the new system

$$[L_p(X)] = [B][X(0) + L_p(E/S)].$$

Remark 2. In a certain number of biological problems the idea of *transit time* is related to a compartment, that is, a *delay r*. Now in the domain of the transformed curve, a delay amounts to multiplying the characteristic function by e^{-pr}, which enables us either to take r into consideration or to calculate it if it is taken as an unknown.

3.3. *Criticism of the method*

The above method is flexible and adapts itself to numerous problems. It enabled us, for example, to determine the coefficients of the model of the kinetics of radioactive iodine [9]. The experimental function known was, in this case, not the quantity contained in a compartment of the model, but a linear combination of the quantities contained in several compartments (thyroid, blood stream, interstitial liquids, hormonal iodine).

Despite such good results, we should, however, point out several inconveniences connected with the very nature of the method:

(i) The Laplace transform upsets the balance of the influence of the "beginning" and the "end" of the curves.

(ii) The choice of the values of the parameter p is more difficult. It has great influence on the convergence of the method. The values should be neither too large, because of the preceding inconvenience, nor too close together. We have not succeeded in setting up a general theory on the optimization of this choice.

4. Principal difficulties encountered and precautions to be taken

4.1. *Number of calculable parameters and experimental data*

Our greatest difficulties are first connected with the number of parameters being too large for the experimental data provided. The models must therefore be simplified. Despite all this, the method remains valuable. Whatever the requirements of the experimenters, it must first be verified that the number of parameters sought is not too large (see Section 3.1.2). However, certain exchanges are known to be impossible in biology. In this case, the corresponding k_{ij} are zero. Quite often, in the matrix A there are many zero elements.

4.2. *Fundamental role of the stage of direct simulation*

Each time we tried to avoid the stage of direct simulation in a new problem we regretted it and, after losing time, we finally were obliged to return to it. It is indispensable for testing the validity of the model by a correct agreement of the result of calculations and of experiment. However, we must not forget that although the "success" of a model is a necessary condition, it is not sufficient. This is why in certain cases simplifications of the model are thus controlled, while preserving a good representation of the phenomenon, for example by associating two compartments [9].

The variations of the parameters produce different effects on the curves. If certain parameters have only negligible influence on the curves available experimentally, it is useless to try to calculate them.

Finally, simulation gives exact values to as many points as required for all the curves which are solutions of the system. The quality of the algorithms of solution of the inverse problem can then be really tested, which is in general much more difficult to do from experimental data.

4.3. *Influence of the parameters and convergence of the calculations*

The convergence of the calculations in the computer is sometimes slow. This difficulty may be explained by a particular structure of the model. This is the case when certain parameters have a closely related influence on the results. If the separate variation of the two parameters produces a practically identical effect, we cannot dissociate them. For problems in which the difference remains slight, we must expect to encounter some difficulties and remain suspicious of the results obtained.

As an example, we give the simulation of the interpretation of the curves of renal isotopic functional exploration [6] in which two coefficients out of ten were calculated with great precision and good convergence, but in which we obtained only mediocre values for the others.

5. Conclusion

The method of models is taking on growing importance in biology and medical functional exploration. It is no longer a simple game, but a true quantitative physiology. However, the major contribution of this modelling is the determination of the parameters of the system. Their calculation from experimental curves implies the resolution of the inverse problem. In all cases the obtained model must be valid, both in normal cases and in pathological cases. We should point out that although pathological models are quite generally superposable on normal models, but with different values for the coefficients, in other cases the pathological models bring in new exchanges in new compartments. This must be thoroughly studied [12, 13].

All this shows the fundamental interest of this mathematical problem, although still incompletely resolved. We are at present trying out other numerical methods to decrease the calculation time and improve the convergence, whatever the initial values of the calculation.

This new aspect of quantitative biology should afford great progress in clinical interpretation of the results obtained. Indeed, quite often (thyroid metabolism, iron metabolism, etc.) the values of the parameters of the model are better correlated with the clinical data than the morphological descriptions normally used to describe the experimental curves, namely by the maximum values at certain privileged moments [7].

We believe that this is a new field of research which is extremely fertile and which, with good collaboration between mathematicians, physicians and biologists, will enable patients to be given better treatment.

Appendix.

Approximation of a function by a linear combination of exponentials

A.1. *Case of a true experimental function (kinetics of iodine)*

Approximation $f = \sum_{i=1}^{3} a_i \exp(-b_i t)$.

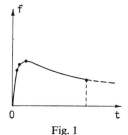

Fig. 1

Table 1

	First group	Second group
b_1	−0.131 767	−0.069 328
b_2	−0.011 282	−0.004 766
b_3	−0.000 026	−0.000 040
a_1	−0.1804	−0.3162
a_2	−0.4750	−0.3933
a_3	+0.6626	+0.7213

Table 2

t_{mn}(min)	Mes	f_1 − Mes	f_2 − Mes
0	0	0.0072	0.0117
5	0.137	−0.0168	−0.0234
10	0.186	0.0035	0.0019
15	0.232	0.0040	0.0109
60	0.420	−0.0006	−0.0008
240(=4 h)	0.590	0.0364	−0.0008
1440(=1 d)	0.680	−0.0419	0.0004
7200(=5 d)	0.540	0.0083	0.0001

A.2. *Case of a simulated function (kinetics of iodine)*

Two approximations: $f_1 = a \exp(bt)$ and $f_2 = a_1 \exp(b_1 t) + a_2 \exp(b_2 t)$.

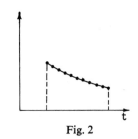

Fig. 2

Table 3

f_1	f_2
$b = -0.00003$	$b_1 = -0.00030$
	$b_2 = -0.00002$
$a = 0.55423$	$a_1 = 0.85449$
	$a_2 = 0.44656$

Table 4

$t(j)$ (min)	Mes	f_1 − Mes	f_2 − Mes
10	0.345	0.0139	0.0002
11	0.333	0.0115	−0.0003
12	0.321	0.0087	−0.0003
13	0.310	0.0057	−0.0001
14	0.300	0.0025	0.0002
15	0.290	−0.0008	0.0004
16	0.281	−0.0043	0.0005
17	0.273	−0.0078	0.0004
18	0.266	−0.0114	0.0001
19	0.258	−0.0150	0.0003
20	0.252	−0.0185	−0.0009

References

[1] P. Berdot, Y. Cherruault, G. Korobelnik, P. Loridan, G. Nissen and F. Tonnellier, Contribution à la résolution de problèmes d'optimisation par des méthodes directes, CNRS no. 68.01.452 (1969).
[2] J. C. Bietry, Etude d'un problème d'estimation des paramètres d'une somme d'exponentielles, Thesis third cycle, University of Nancy (1971).
[3] P. Faurré and E. Irving, L'identification des systèmes linéaires par l'algorithme de B.L. Ho, Chim. Ind. (Paris), Génie Chim. 103 (1970) 1–8.
[4] J. Legras, Méthodes et techniques de l'analyse numérique (Dunod, Paris, 1971).
[5] J. Martin, Les modèles de métabolisme et l'analyse compartimentale, Bull. Inform. Méd. (3) (1969) 101.
[6] J. Martin and C. Monot, Analysis of a mathematical model of renal function. Numerical adjustment of parameters on a computer by reference to isotopic néphrogram curves using Hippuran I^{131}, in: L. Timmermans and G. Merchie, eds., Radioisotopes in the Diagnosis of Diseases of the Kidneys and the Urinary Tract (Excerpta Medica, Amsterdam, 1969) 508–520.
[7] J. Martin and C. Monot, Formulation simplifiée des sytèmes biologiques d'échange et ajustement de leurs paramètres à partir d'une ou plusieurs courbes expérimentales, in: Proc. 6th Congr. A.F.I.R.O., vol. 6 (1967) 105–136.
[8] J. Martin and C. Monot, Simulation analogique de la fonction thyroïdienne, Ajustement de certains coefficients d'échange entre compartiments à partir de la courbe de fixation cervicale de l'iode radioactif, Symp. de l'Informatique Médicale, Toulouse, vol. 2 (1967) 39–66.
[9] J. Martin, J. Finas, F. Georges and J. Robert, Etude de l'enregistrement cervical et thyroïdien dans les 15 minutes qui suivent l'injection intra-veineuse d'iode radioactif: taux de vascularisation cervicale, clearance iodée, Ann. Radiol. 8 (1965) 665–683.
[10] J. Martin, C. Monot and O. Foucaut, Quelques difficultés de l'analyse mathématique des échanges entre compartiments, in: Proc. Internl. Meeting on Computers in Radiology, Brussels, 1969 (Karger, Basel, 1970) 423–427.

[11] C. Monot, Ajustement numérique d'un système différentiel linéaire à l'aide de la transformation de Laplace, Thesis third cycle, University of Nancy (1967).
[12] C. Monot, J. Martin, Y. Najean and C. Dresch, Problèmes posés par la simulation de la cinétique du fer, *Journées d'Informatique Médicale, Toulouse* (1969) 219–241.
[13] C. Monot, J. Martin, Y. Najean, C. Dresch and A. Faille, Simulation de la cinétique du fer radioactif dans les cas pathologiques (problèmes mathématiques), *Journées d'Informatique Médicale, Toulouse*. vol. 2 (1972) 183–199.
[14] D. H. Parsons, Biological problems involving sums of exponential functions of time: a mathematical analysis that reduces experimental time, *Math. Biosci.* 2 (1) (1968) 123–128.
[15] D. H. Parsons, Biological problems involving sums of exponential functions of time: an improved method of calculation, *Math. Biosci.* 2 (1) (1970) 37–47.
[16] S. Rubinow and A. Winzer, Compartment analysis: an inverse problem, *Math. Biosci.* 11 (3/4) (1971) 203–247.
[17] C. W. Sheppard, *Basic Principles of the Tracer Method* (Wiley, New York, 1962).
[18] L. Schwartz, *Etude des sommes d'exponentielles* (Hermann, Paris, 1959).
[19] G. Vallée, J. P. Belin and G. Beaugas, Programme approchant par une courbe pluriexponentielle des données expérimentales discrètes obtenues chez l'homme par détection radioactive, *Symp. sur l'Informatique Médicale et la Gestion Automatisée d'un Hôpital, Toulouse* (1968) 165–178.

DISCUSSION

Groth:

I would like to comment on the paper we just heard with regard to numerical versus analytical methods. My experience is that numerical integration methods, e.g. the well-known Runge–Kutta methods, have apparent advantages both with regard to (i) modelling of compartmental systems and to (ii) parameter estimation.

From the practical programming point of view, numerical methods are more flexible and easier to apply when dealing with complex systems and when trying several alternative model structures in the initial phase of the modelling work (cf. e.g. Groth et al.[1]). Furthermore, analytical methods often put restrictions on the experimental procedure (constant plasma level, etc.), disadvantages which can easily be avoided by the use of numerical methods (see e.g. Arturson et al.[2]). For the biomedical researcher, with often little experience of sophisticated mathematical problems, general purpose simulation programs (for a review see e.g. Hentsch[3]) after new valuable possibilities, especially in combination with interactive time-sharing computer systems.

[1] T. Groth et al., Computer simulation of ferrokinetic models, *Comp. Progr. Biomed.* 1 (1970) 90–104.

[2] G. Arturson et al., The exchange kinetics of sodium, potassium, sulphate ions and albumin in the dog's heart, *Acta Phys. Scand.*, submitted.

[3] W. Hentsch (Oldenburg, München, 1969).

Concerning economy of computing, this might be a local problem, but the present development in computer manufacturing, with decreasing prices, in particular for the central processing units, is in favour of numerical methods.

The procedure of fitting sums of exponentials for the experimental data give (rather large) errors in intercepts and slopes, errors which propagate (and are magnified) in the calculations of the rate constants from algebraic formulas. On the other hand, a *direct* least square fit of simulated curves, as calculated by numerical integration methods, by the use of efficient non-linear parameter search routines (e.g. Powell[4]) gives more confident estimates of the rate constants (see Tengström[5]).

The problem not to use too many parameters (over-complex models) to fit experimental data is a general problem in model-building, which should be checked by calculation of the confidence of the estimated parameters. The adequacy of a model, which in addition to compatibility to experimental data also requires that the parameters are well-determined, may advantageously be illustrated by plotting (multi-dimensional) confidence regions. The shape and the "dynamics" of these regions in the parameter space are of good help in the judgment of adequacy and in the design of optimal experiments (e.g. the spacing of points in the experimental domain with due consideration to measuring errors).

Martin to Groth:

We completely agree on the advantages of numerical calculation and computer aid, on both points of view, modelling and resolution. Particularly, we emphasize the danger of exponential fitting, except if it is only an intermediate step of the calculation, as in our case.

Determination of variance of coefficients is very important. Dr. Valleron has developed a good method for this. In fact, the real problem is not numerical method but the adequacy of modelling and the quality (and number!) of data.

Validity of model may be suggested by a simulation of all the cases and comparison with reality, both in normal and pathologic cases; is there the same model?

Practically, each hypothesis may be recognized and discussed, i.e., a complete analysis of situation.

On the other side, it is often difficult to obtain of experimenters as many data as necessary. Among our methods, some give a great weight to initial

---

[4] M. J. D. Powell, *Comp. J.* 7 (1965) 303–307.

[5] G. Tengström, Clearance of I^{125}, in: *On the Analysis of Data from Single Injection Experiments* (to be published).

points, so we often ask to have initial registration of phenomena. The chronological distribution of measures is also difficult to optimize.

We ought to convince experimenters that computer calculation cannot supply data.

Reich:

The main point does not seem to be which numerical method should be used (although I agree with Dr. Groth that direct fitting is better than peeling off the exponentials), but whether or not there exists a unique solution of the "inverse problem". If not, then no method will give satisfactory results. In fact, Dr. Monot's results indicate that very often multiple parameter sets exist which describe the data equally satisfactory. A computing routine should be able to exclude or recognize such pitfalls.

Martin to Reich:

I don't agree completely with you. In fact, the solution of inverse problem depends on the number of data, i.e., the number of independent experimental curves ($n - 1$, if n compartments). If you haven't got all these results, sometimes you have a multiple parameter set. You ought then to put complementary conditions if you known criteria. In one example (heterozygotes of phenylketonuria), we have a solution with negative k_{ij}, and this has no physiological significance.

The diapositive gives you an alternative convergence on each of these solutions.

Table 1

Phenylketonuria – A case of convergence with alternation of results

No iteration	S	k_1	k_2	k_3	$\sum_i |F_i - \text{Mes}_i|$
Without initial constraints	0.8041	0.0477	0.0152	0.0145	0.0269
2	0.2193	0.4430	−0.0065	0.2946	0.4171
3	0.5986	0.1145	0.0030	−0.0007	0.2901
4	0.2254	0.4932	0.0210	0.0103	0.0832
5	0.5964	0.1145	0.0010	−0.0060	0.3255
6	0.2254	0.4928	0.0209	0.0104	0.0831
7	0.5964	0.1145	0.0010	−0.0060	0.3255
8	0.2254	0.4927	0.0209	0.0104	0.0831
9	0.5964	0.1145	0.0010	−0.0060	0.3254
10	0.2254	0.4927	0.0209	0.0104	0.0831
11	0.5964	0.1145	0.0010	−0.0060	0.3255

On a theoretical point of view, I would like to have a theorem for conditions of uniqueness of solution. It is the mathematician's job.

X to Martin:
Couldn't you replace, by programming, each unacceptable negative coefficient by zero?

Monot to X:
We try to do it, with bad results. In fact, it is a curious calculation phenomenon like this you can see on Table 2.

Table 2

Phenylketonuria – Two solutions of which only one is acceptable

Iteration no.	S	Set I			Set II			$\sum_i F_i - M_i$
		k_1	k_2	k_3	k_1	k_2	k_3	
Without initial constraints	0.412	0.2393	0.0258	0.0165				0.0031
2	0.166	0.7724	0.0552	0.0846	0.0484	−0.4859	1.1349	0.0718
3	0.357	0.2666	0.0061	0.0153	0.0119	−0.0662	0.3423	0.0031
4	0.238	0.4313	0.0080	0.0154	0.0096	−0.2405	0.6855	0.0013
5	0.288	0.3455	0.0063	0.0150	0.0113	−0.1051	0.4607	0.0019
6	0.262	0.3864	0.0071	0.0152	0.0106	−0.1546	0.5527	0.0016
7	0.274	0.3661	0.0067	0.0151	0.0110	−0.1276	0.5047	0.0016
8	0.268	0.3760	0.0069	0.0152	0.0108	−0.1400	0.5227	0.0016
9	0.271	0.3713	0.0069	0.0152	0.0109	−0.1333	0.5166	0.0015
10	0.270	0.3735	0.0069	0.0152	0.0109	−0.1367	0.5216	0.0015
11	0.270	0.3725	0.0069	0.0152	0.0109	−0.1353	0.5190	0.0015

We have no explanation for this alternance of two sets of coefficients, with sequence of positive and negative values for k_2.

Valleron to Monot and Martin:
In our laboratory, with J. Y. Mary, we have developed a simple Monte Carlo method which makes it possible to compute the distribution of the unknowns k_{ij} if the distribution of the experimental values is known or estimated. In this method instead of using the original set of experimental points, we create many sets of "pseudo-experimental" points: each pseudo-experimental point is chosen randomly in the confidence interval of the original value (for instance, if at time t our observer experimentally N counts of radioactivity, the pseudo-experimental points will be drawn from the Poisson distribution of mean N).

Using a first set of pseudo-experimental points, we obtain a first set of solution for the k_{ij}. Using a second set of pseudo-experimental points, we obtain a second set of solution for the k_{ij} and so on. After iteration, we obtain the distribution of any k_{ij}. The method may be used to check whether or not a coefficient is significantly different from 0.

This approach is a supplementary step in the problem of the estimation of the parameters of a model. It may be used with numerical or analytical values. Its cost is obvious: it needs to use the research procedure as many times as we have generated sets of pseudo-experimental values.

RESISTANCE OF CROPS TO DISEASES: A GAME-THEORY MODEL

J. PEŠEK

Department of Biometrics, Academy of Agriculture, Hrušovany u Brna, Czechoslovakia

A couple of years ago, I heard of an excellent example of the failure of statistical inference that has bothered me ever since. There was a certain pond. One day, 20 fishes were caught in it and the next day 10. How many would one expect to catch on the third day? The speaker was so kind as to present the maximum-likelihood solution, but I suspect he was pulling our leg, because I do not believe that the problem in fact has any solution, at least not in the terms in which it is expressed.

One straight-forward approach is to argue that fishing is a matter of luck and our values of 20 and 10 are two samples from a Poisson distribution with a mean that we should estimate as about 15. It would be reasonable for someone to expect that number of fishes to be caught on the third day.

It would also be reasonable to assume that on the first day one caught Np fishes out of the N originally in the pond, and on the next day $N(1-p)p$ out of the $N(1-p)$ left. On the third day one would expect $N(1-p)^2 p$ to be caught, i.e. five.

We have, however, reckoned without the Detective Inspector, who is convinced that poachers took as many fish as they could on the first day and came back for the rest on the second. If he is right, there will be none to catch on the third.

Clearly we have no way of resolving the conflict of opinion except by going fishing at the third day and seeing what happens. If even one fish is caught, the Detective Inspector's conjecture is badly shaken. If the number of fishes exceeds ten, it is difficult to believe that the population is being depleted, as the second person thought. If the number should exceed 25, we must suspect that we are confronted by a phenomenon that we just do not understand.

When dealing with data from phytopathological studies we are confronted with a similar situation. The usual statistical inference does not yield the conclusions we are asked for. System analysis, namely the theory of games, may be thus helpful in the study of a relation between hosts and pathogens.

In phytopathology, the main components of the system are the host crop, the pathogens or parasites (often a vector), and the microclimate. There may be two ways how to control plant diseases: either to modify the microclimate by use of fungicides, or to develop varieties genetically resistant to diseases. The latter is more expensive, but perhaps more humane since it does not damage the environment.

The theory and practical uses of multilineal varieties in self-pollinated species have been developed since 1957 by Putt for rust resistance in sunflowers, by Nobel Prize winner Borlaug for stem rust resistance in wheat, by Gibler for stem rust and yellow rust resistance in wheat, and also by Browning and Frey for crown rust resistance in oats. The significance of genetic diversity was greatly emphasized by Suneson in 1960. Part of the Suneson's data are shown in Table 1.

Table 1

Stem rust development in ONAS 41, ONAS 43, and in their mixture at Davis, Calif. (1957).

Observation	Number of rust pustules		
	ONAS 41 (susceptible)	ONAS 53 (resistant)	ONAS 41: ONAS 53 (1:3 mixture)
Stem rust April 1	2	0	0
May 1	5–50	0	0
May 21	10–90	0	0–20
May 28	90	0	0–60
Heading	April 20	April 20	April 20
Ripening	June 1	June 6	June 6
Height (cm)	115	115	115
Yield (q/ha)	23	54	50

According to his experimental results, the 1:3 mixture of the susceptible variety ONAS 41 and the resistant variety ONAS 53 did not yield significantly lower than the ONAS 53. He pointed out that the moderate decrease of resistance and yield was due to dilution of buffering protection for ONAS 41 plants in the mixture.

For the achievement of the required genetic diversity, two types of breeding procedures may be proposed: The first is an accumulation of different kinds of resistance genes into a pure line variety, and the other is a mechanical mixture of them in a variety. A multilineal variety is superior in several points if compared with a pure line. For example, it can be reconstructed within a fairly short period, when any of the components becomes useless due to its change in susceptibility to new races of pathogens. Of course, in this case it is also necessary to develop donor resistant lines in advance. However, breeding seems to be more simple and requires considerably less time and labour as compared with conventional cross-breeding of pure line varieties.

Now, we are faced with a new problem: what proportion should be assigned to each constituent of resistance genes within a multilineal variety? Relative to this problem, three types of construction could be assumed:

(i) the more different resistance genes, the better for a lasting effect;

(ii) a fairly larger number, but not too many, of resistance genes would be desirable;

(iii) a mechanical mixture of resistance genes might not be suitable, but a subsequent exchange of them, one by one or a few by a few, might be more effective.

The first type is based on the concept that the predominant physiologic races of a parasite can hardly be predicted in advance. Therefore, a greater variability would assure an enduring effect of disease resistance.

The second type is based on the assumption that predominant races of the parasite can be predicted in advance, even though approximately.

The third type, that is a subsequent exchange of the resistance gene or genes, is based on the same assumption as the second. However, a mixture of the second type might be accompanied by a greater complexity of race constituents, since several resistance genes are used at one time in a mixture. Consequently, when all of the resistance genes have lost their effects, the immediate exchange with new resistance genes would be extremely difficult due to the lack of new resistance gene sources.

Anyway, the common expectation with these breeding procedures is that the greater variability in host populations would assure an enduring effect with respect to disease resistance. Now, from these concepts and experiences it may be deduced that there exists the following situation between host and parasite populations:

(1) A parasite cannot exist without its host. This means that it seems impossible for the host to completely eradicate parasites.

(2) In this situation, corresponding to the structure of the resistance and susceptibility genes in the host population, the ratio of the virulent and avirulent genes in the parasite population will have an appropriate change so as to make a long term equilibrium status between the host and parasite populations.

(3) For breeding purposes, the level of the equilibrium should be kept as low as possible with respect to the damage by disease.

In this situation, "equilibrium" means that the maximum loss for the host should be balanced with the minimum violence of the parasite to ensure the existence of the parasite population. Therefore, this relationship between host and parasite could be replaced with the competitive situation in which the gain of the one side (A) corresponds to the loss of the other, say B. A will use its various strategies to get the maximum gain, while B will use its possible strategies to minimize its loss. For this reason, the competitive relationship seems to be replaced with the situation in game theory of operations research. The resistance genes of the host and the races of the parasites are the two players. It is assumed that the response of the resistant genes to any races can be quantitatively expressed by appropriate values.

Now gene A shows a resistance grade, say 3 and 0, to races N1 and C1, respectively, where high resistance is graded 0 and 1, 2, 3, . . . show the increase of susceptibility to disease (Table 2). The symbols $x_1, x_2, \ldots, x_n$ in the mixture represent the frequencies of each resistance gene in the host. If the parasite population consists of either N1 race or C1 race only, the response should be 3 or 0. If the population of the parasite is a mixture of those two races, then the response of the host is proportional to the rate of the two races. As the second situation is assumed the one in which gene B is additionally included in the host whose response is 1 and 3 to parasite races N1 and C1, respectively. If we may assume that competitive features between host and parasite might be compared with a play in a game of operations research, the optimum mixture of the resistant genes would cause the maximum loss of the host, and the expected loss or response is 1.8, which is estimated as the game's value. This means that the loss of the host should be less than 1.8 at long term, and also the gain of the parasite may be expected to be 1.8, anyhow, at the minimum.

Now, if another gene, say D, is included in the host (Case 5), it would be expected that the game's value is slightly decreased to 1.7 from the value 1.8 in Case 3 or 4. Furthermore, as in Case 6, resistance genes, say I and N, are used with gene A, the game's value is expected to be largely reduced to 0.75, which means that the equilibrium is maintained at a more desirable level than in Cases 3, 4 and 5. In this situation, genes A, I and N should be mixed in a rate of $\frac{1}{4}$, $\frac{1}{4}$ and $\frac{1}{2}$, respectively, in the host. In such a way, our expectation is that the constituents of the parasite could be regulated, to some extent, by deliberate mixing of different kinds of resistance genes.

However, we have the following problems regarding the application of game theory to practical use:

(i) It is necessary to have information in advance on the response of each gene in the host to each of the races in the parasite.

(ii) Positive or negative cooperation among the races of the parasite, if any, are not considered in this situation.

Table 2

Hypothetical reaction to disease of a host population in a mixture of resistance genes.

Case	Resistance genes in a host population	Races in a parasite population		Mixture $x_1\ x_2\ x_3\ ...\ x_i\ ...\ x_n$	Expected reaction (g)
		N1	C1		
1	A	3 or	0	1 - - ... - ... -	3 or 0
		N1	C1		
2	A	3	0	1 - - ... - ... -	$3 \geq g \geq 0$
3	A B	3 1	0 3	$\frac{2}{5}$ $\frac{3}{5}$ - ... - ... -	1.8
4	A B C	3 1 4	0 3 5	$\frac{2}{5}$ $\frac{3}{5}$ - ... - ... -	1.8
5	A B D	3 1 0	0 3 4	$\frac{4}{7}$ 0 $\frac{3}{7}$ ... - ... -	1.7
6	A B . I . N	3 1 . 2 . 0	0 3 . 1 . 1	$\frac{1}{4}$ 0 0 ... $\frac{1}{4}$ ... $\frac{1}{4}$	0.75

(iii) Neither migration of new races nor mutated races in the parasite are considered here.

(iv) From the practical point of view in breeding, the effect of the mechanical mixture at one time should be compared with that of the subsequent exchange of resistance genes.

This paper started with system analysis, but dealt only with one of its elements – simulation of the composition of the resistance genes in multilineal varieties. Its purpose was to turn your attention to the field in which applied mathematics has not yet found a proper place. It is my firm belief that in plant phytopathology, mathematicians and pathologists must work hand in hand toward a deeper understanding of the epidemiological process, and possibly toward the prognosis of the various forms of partial resistance and disease warning services.

THE INFORMATION CONTENT OF KINETIC DATA

J. G. REICH and I. ZINKE

Zentralinstitut für Molekularbiologie, Bereich Methodik und Theorie, Akademie der Wissenschaften der D.D.R., Berlin-Buch, D.D.R.

Many biological experiments can be described by a kinetic model equation of the form

$$y = f(x,p) + \varepsilon,$$

where x is a d-vector of input quantities ($x \in X \subseteq \mathbf{R}^d$), p is a n-vector of kinetic parameters ($p \in P \subseteq \mathbf{R}^n$), and y is an observable scalar, whereas ε describes the stochastic influence. It follows that empirical determinations of y can be of limited experimental accuracy only. x may in practice be a vector of concentrations of influencing ligands, and y the enzymic activity. But the model is valid also in other contexts (exponential decay functions, etc.). f is assumed to be a continuous real function of its arguments (x, p). Then, given a set of experimental data (y_k, x_k), one can define a minimum distance estimate $\hat{p}$ of p:

$$\phi(\hat{p}) \leq \phi(p), \quad p \in P \quad \text{(least squares)},$$

$$\phi(p) = \sum_k (w_k r_k)^2 \quad \text{(sum of squares)},$$

$$r_k(p) = f(x_k, p) - y_k \quad \text{(residuals)},$$

In the present paper the following problems which arise in the context of such kinetic models have been treated:

Problem 1 (compatibility of model and data). Given (y_k, x_k), f, $\hat{p}$. Are there systematic deviations between model and observed data?

Problem 2 (design satisfactory?). Given (y_k, x_k), f, $\hat{p}$. Is the experimental set up good enough to estimate the parameters in a unique manner?

Problem 3 (redundancy of model with respect to parameters). Given f, region P. Strength and mutual independence of the influence of the parameters on the model.

Problem 4 (parameter region). Given $(y_k, x_k), f, \hat{p}$. In which region is the random variable p to be expected? (The region is the analogue of the standard deviation in univariate statistical analysis.)

The detailed presentation of the solutions to these problems as proposed by us would require the amount of a full length paper. Therefore the general line will be sketched only.

Problem 1. This is tackled by investigation of the homogeneity or inhomogeneity of the distribution of r_k in the design space.

Problem 2. Transformation to the principal axes of the information matrix:

$$\hat{J}_{ij} = \frac{1}{m} \sum_k \frac{\partial f(x_k, p)}{\partial p_i} \frac{\partial f(x_k, p)}{\partial p_j} w^2(x) \bigg|_{p=\hat{p}}.$$

Problem 3. Transformation to the principal axes of the model matrix:

$$J_{ij} = \int_X w^2(x) \frac{\partial f(x, p)}{\partial p_i} \frac{\partial f(x, p)}{\partial p_j} dx \bigg/ \int_X w^2(x) \, dx.$$

Problem 4. Inversion of J (similar to the classical linear estimation problem).

This analysis leads to three empirical quantities of a model, the information measure, the sensitivity measure, and the redundancy measure. The contribution of each parameter to these coefficients can be evaluated. Practical results are statements on the optimization of the design and on possible or impossible parameter estimation.

Reference

J. G. Reich, G. Wangermann, M. Falck and K. Rohde. A general strategy for parameter estimation from isosteric and allosteric-kinetic data and binding measurements, *Eur. J. Biochem.* 26 (1972) 368–379.

COMMENTS

Groth:

First I want to congratulate Dr. Reich for the interesting presentation on the basic problems of modelling. I am working along the same lines and I hope to have the opportunity to show you some examples during the panel discussion in the afternoon. Just now I want to make the remark that the estimation situation may be properly visualized by plotting likelihood contours.

In this connection it might also be relevant to note the possibility of using this as a tool of designing "optimal" experiments by studying the dynamics of the likelihood contours.

A MATHEMATICAL MODEL OF CELLULAR DIFFERENTIATION

Bl. SENDOV

Mathematical Institute, Bulgarian Academy of Sciences, Sofia, Bulgaria

and

R. TSANEV

Biochemical Institute, Bulgarian Academy of Sciences, Sofia, Bulgaria

1. Introduction

The purpose of our modelling was to check the plausibility of a hypothesis concerning the mechanism of cellular differentiation, i.e. the emergence of different kinds of cells from a single embryonic cell as a result of many divisions. The elucidation of the mechanism of cellular differentiation is one of the most important in modern molecular biology. Our feeling is that this is the only way of understanding also the mechanism of carcinogenesis, which can be considered as an abnormal differentiation. It is obvious that such a type of modelling could lead to a definitive result of a negative character only. A hypothesis can be fully rejected in case the corresponding model does not reflect the basic characteristics of the modelled object. On the other hand it is not a proof that the hypothesis is true if the mathematical model reflects very satisfactorily the basic features of the modelled object. This can only make the hypothesis a more reliable candidate for direct checking through real experiments. This method gives the possibility rapidly to check and to perfect hypotheses due to the great flexibility of mathematical modelling, compared to the possibilities of carrying out real experiments.

Our hypothesis of the mechanism of cellular differentiation is partly based on a number of facts already well established in molecular biology [6].

Here the mathematical model only will be described and some results of the computer experiments will be shown.

2. The model of eukaryotic cells [7]

The model of an eukaryotic cell should include as a basic control circuit a metabolic pathway controlled by an operon and leading to the synthesis of specific products which can influence the same or other metabolic circuits.

Each element of such a circuit represents a process which can be described by mathematical equations. The activity of an individual cell is controlled by a fixed number of such metabolic circuits which can be functionally interconnected on the basis of repression, repressor modification and deblocking [see 6]. The whole set of metabolic circuits interconnected in this way forms a complex genetic net which determines the behaviour of the individual cell. Therefore, the model of a multicellular eukaryotic system will be characterized by the following elements:

(1) A genetic net of m interconnected control circuits. The functional interactions between these circuits can be fixed by means of three matrices (of repression, repressor modification and deblocking).

(2) A multicellular configuration fixed by the contacts between cells and the permeability of the cell membrane for different cell products.

The metabolic pathway of each operon is described by the following variables, which depend on the time t:

$C_j^i(t)$ = concentration of mRNA,
$X_j^i(t)$ = concentration of programmed ribosomes,
$P_j^i(t)$ = concentration of different proteins,
$R_j^i(t)$ = concentration of repressors,
$Y_j^i(t)$ = quantity of the deblocking protein in the operon.

The upper indices of all variables refer to the different cells, while the subscripts refer to the corresponding operon of all cells.

It is assumed that only $R_j^i(t)$ can diffuse throughout the cell membrane.

The interrelations between operons in a single cell are defined by three matrices: the matrix of repression $[\sigma_{ij}]$, the matrix of deblocking $[\kappa_{ij}]$ and the matrix of repressor modification $[\alpha_{ij}]$. In this report we shall consider the first two matrices only. The element σ_{ij} defines the effect of the repressor produced by the ith operon on the jth operon. The element κ_{ij} defines the deblocking effect of the protein produced by the ith operon for the deblocking of the jth operon.

The state of each operon is defined by two binary variables $\xi^i_j(t)$ and $\eta^i_j(t)$:

$$\xi^i_j(t) = \begin{cases} 0 \text{ (repressed)} & \text{if } \sum_{k=0}^{m} \sigma_{kj} R^i_k(t) > A_j, \\ 1 \text{ (derepressed)} & \text{if } \sum_{k=0}^{m} \sigma_{kj} R^i_k(t) \leq A_j, \end{cases} \quad (1)$$

$$\eta^i_j(t) = \begin{cases} 0 \text{ (blocked)} & \text{if } y^i_j(t) < \tfrac{1}{2}, \\ 1 \text{ (deblocked)} & \text{if } y^i_j(t) > \tfrac{1}{2}. \end{cases} \quad (2)$$

The constants $A_j (j = 0, 1, 2, \ldots, m)$ are the threshold levels of repression [see 3].

In order to follow the state of the genetic net of a cell, we should define the interrelations between the variables characterizing this cell. On the basis of considerations discussed in detail in our papers [1–7], the following basic relations have been derived:

$$\frac{dC^i_j(t)}{dt} = a_j \frac{\xi^i_j(t)\,\eta^i_j(t)}{1 + \sum_{k=0}^{m} \sigma_{kj} R^i_k(t)} - C^i_j(t), \quad (3)$$

$$\frac{dX^i_j(t)}{dt} = b_j (C^i_j(t) - X^i_j(t)), \quad (4)$$

$$\frac{dP^i_j(t)}{dt} = c_j (X^i_j(t) - P^i_j(t)), \quad (5)$$

$$\frac{dR^i_j(t)}{dt} = f_j (X^i_j(t) - R^i_j(t)) - \lambda_j (R^i_j(t) - R^e_j(t)). \quad (6)$$

The variable $R^e_j(t)$ defines the concentration of the corresponding repressor out of the cell and λ_j the permeability of the cellular membrane. The evaluation of $R^e_j(t)$ depends on the geometrical configuration formed by the cells and the contacts between the cells.

Eq. (3) shows that the jth operon O^i_j of the ith cell will produce mRNA if $\xi^i_j(t) = 1$ and $\eta^i_j(t) = 1$. If not, the concentration $C^i_j(t)$ will tend to 0 exponentially. The production of all other substances controlled by the operon O^i_j depends on $C^i_j(t)$.

$Y^i_j(t)$ is a stochastic variable and can have the values 0, $\tfrac{1}{2}$ and 1. If $Y^i_j(t) = 1$, it remains 1. If $Y^i_j(t) < 1$, then the value of $Y^i_j(t)$ is increased by $\tfrac{1}{2}$ in the time interval $[t, t + \Delta t]$ with a probability proportional to $\Delta t \sum_{k=0}^{m} \kappa_{kj} P^i_j(t)$. Depending on the values of $\xi^i_j(t)$ and $\eta^i_j(t)$, the operon O^i_j at a given moment t can be in one of the four different states shown in Table 1. Only the latter state is active in transcription.

Table 1

State	$\xi_j^i(t)$	$\eta_j^i(t)$
Blocked repressed	0	0
Blocked derepressed	1	0
Deblocked repressed	0	1
Deblocked derepressed	1	1

Taking into consideration (1) and (6), one can see that the state of repression and derepression of a given operon can be changed depending on the changes in the concentration of some repressors. Such changes can also be caused by outer influences due to the fact that in our model repressors are supposed to diffuse through the cellular membrane. This may not be the real case, but formally such an assumption is equivalent to a situation where a low molecular weight effector can diffuse and modify the repressor (repressor modification).

It is seen that in our model the activity of the genome is controlled by two different mechanisms: a mechanism of blocking–deblocking, based on an irreversible binding of proteins to the chromatin, and a mechanism of repression–depression based on a reversible binding of repressors.

It is evident both from the definition of $Y_j^i(t)$ as a stochastic variable which for the time being can only increase and from the definition of $\eta_j^i(t)$ (eq. (2)), that an operon can change only from a blocked to a deblocked state.

The reverse change – from a deblocked to a blocked state – can occur only as a result of a mitotic division.

The mechanism of cellular division in our model is realized in the following way (see [3]): We suppose that the cell contains a group of operons (mitotic operons) which are repressed by other operons during the period of functional activity of the cell. Their derepression makes the cell enter the mitotic cycle. One of the mitotic operons has been selected to command the entering of the cell into the mitotic cycle. For example, if we select for this purpose the operon O_0^i, then when the concentration of the regulatory protein $P_0^i(t)$ reaches the threshold level B_0 ($P_0^i(t) \geq B_0$), the ith cell enters the mitotic cycle and then divides. The mitotic cycle in the model can be described with different details. The essential point is that each cellular division increases the number of cells by one. According to the accepted cellular configuration (arrangement of cells), the place of each new cell and its contacts with the other cells should be determined. The two daughter cells at the very moment after division possess all substances in concentration equal to the respective concentrations of the mother cell. The quantity $Y^i_j(t)$ of the mother cell can be distributed in the ith and $(i+1)$st daughter cells according to two different rules:

(i) If $Y^i_j(t) = 1$, then $Y^i_j(t) = Y^{i+1}_j(t) = \frac{1}{2}$.
If $Y^i_j(t) = \frac{1}{2}$, then $Y^i_j(t) = \frac{1}{2}$, $Y^{i+1}_j(t) = 0$,
or $Y^i_j(t) = 0$, $Y^{i+1}_j(t) = \frac{1}{2}$.
If $Y^i_j(t) = 0$, then $Y^i_j(t) = Y^{i+1}_j(t) = 0$.

(ii) Independently of the value of $Y^i_j(t)$ in the mother cell we can assume that $Y^i_j(t) = Y^{i+1}_j(t) = 0$ for the daughter cells immediately after division.

Rule (i) corresponds to the conservation of the deblocking proteins, while rule (ii) corresponds to their destruction during the mitotic cycle.

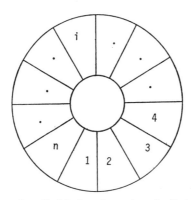

Fig. 1. Cross section of a cylindrical configuration of cells (model "Kylindros").

In both cases an operon that has remained blocked after division can be deblocked soon after if a corresponding deblocking protein is available in the cell. If such a deblocking protein is lacking or its concentration is very low, the corresponding operon will remain blocked.

It is clear that after and only after division an operon that has been deblocked in the mother cell may become blocked in one of the daughter cells. This is in fact the mechanism leading to differentiation of cells according to our model.

With an appropriate choice of the matrices $[\sigma_{ij}]$ and $[\kappa_{ij}]$ it is possible to start the model experiments with one single cell, in which some of the operons are deblocked. This cell may undergo a fixed number of divisions, and as a result the pattern of deblocked operons can be changed, i.e., differentiation will occur and a state can be reached where no more divisions take place.

To facilitate our task, we have accepted a simpler configuration of cells representing an endless cylinder with an axial hole and only a cross section of this cylinder was considered; Fig. 1 (model "Kylindros" [6]).

3. Computer experiments

In order to test the possibilities of the molecular mechanism of our model to reproduce the phenomena of cytodifferentiation, some quantitative evaluations were necessary. For this purpose we tried to examine the development of a single "Kylindros" cell which was activated to divide and in which all or some of its operons were inactive at the moment of starting. In such a model there was no possibility of development unless some initial information were present in the cytoplasm to initiate the genetic program. This means that a specific deblocking protein (or proteins) should be present or synthesized in order to deblock the first operon(s) in the chain leading to further development. Such a requirement is consistent with the finding of stored mRNA in the cytoplasm of animal eggs and in plant seeds (see [6]).

To realize our computer experiments, it was necessary to fix the number of operons, to select genetic nets out of the extremely large number of possibilities described by the corresponding matrices $[\sigma_{ij}]$, $[\kappa_{ij}]$ and $[\alpha_{ij}]$, and to choose numerical values for all parameters of the system. Due to the limitations in the computer time and main storage (Minsk 22), a model consisting of only eight operons in each cell was studied, only seven of them being used in the present experiments. The maximum number of cells which could be followed during the development of this system was 81. In this model each operon may be interpreted not only as an individual operon but also as a group of interrelated operons which may have its own subprogram of development fixed by the same mechanism as the general program. In other words, the activation of the operons in the scheme also may be considered as the starting point of different subprograms.

The complexity of the system and the great number of parameters whose numerical values are unknown required a great number of preliminary computer experiments to make a reasonable choice of a working set of parameters, i.e. parameters of a cellular system which comes to a steady state after a certain number of cellular divisions and differentiations have taken place.

The initial program, imitating "stored" mRNA or the presence of a deblocking protein, was realized by a special compartment of the computer program which could maintain a given concentration of a variable for a fixed period of time. In terms of biochemical events this period reflected a fixed half-life time of the stored mRNA, a fixed degradation rate of the corresponding deblocking protein and a fixed membrane permeability of the cell.

Three groups of computer experiments were carried out. In the first group the influence of the duration of the initial "stored" program was studied, in the second group the influence of the number of initial cells and of the threshold values A_j, and in the third group the influence of contacts between cells. The following numerical values of the parameters were selected as common to the three groups:

Table 2

Parameters \ operon i	$i=0$	$i=1,2\ldots 6$	$i=7$	
$a_i = b_i = c_i$	0.10	0.20	0	$B = 0.98$
f_i	0.20	0.10	0	

3.1. First group

The matrices of repression and deblocking were selected as follows:

$$[\sigma_{ij}] = \begin{bmatrix} 0 & 0.20 & 0.20 & 0 & 0.20 & 0 & 0.20 & 0.01 \\ 0.20 & 0 & 0 & 0 & 0.20 & 0 & 0.20 & 0 \\ 0 & 0 & 0 & 0.20 & 0 & 0 & 0 & 0 \\ 0 & 0 & 0 & 0 & 0 & 0 & 0 & 0 \\ 0.20 & 0.20 & 0.20 & 0.20 & 0 & 0.20 & 0 & 0 \\ 0 & 0 & 0 & 0 & 0 & 0 & 0 & 0 \\ 0.20 & 0.20 & 0.20 & 0 & 0.20 & 0.20 & 0 & 0 \\ 0 & 0 & 0 & 0 & 0 & 0 & 0 & 0 \end{bmatrix},$$

$$[\kappa_{ij}] = \begin{bmatrix} 0.99 & 0 & 0 & 0 & 0 & 0 & 0 & 0 \\ 0.99 & 0.99 & 0 & 0 & 0 & 0 & 0 & 0 \\ 0 & 0 & 0.40 & 0.99 & 0 & 0 & 0 & 0 \\ 0 & 0 & 0 & 0 & 0.99 & 0 & 0 & 0 \\ 0.99 & 0 & 0 & 0 & 0.99 & 0.99 & 0 & 0 \\ 0 & 0 & 0 & 0 & 0 & 0 & 0.99 & 0 \\ 0.99 & 0 & 0 & 0 & 0 & 0 & 0.99 & 0 \\ 0 & 0 & 0 & 0 & 0 & 0 & 0 & 0 \end{bmatrix}.$$

The following threshold values of the parameters A_j were chosen:

j	0	1	2	3	4	5	6	7
A_j	0.2350	0.2360	0.2400	0.0005	0.2700	0.0005	0.3600	1.000

We started the experiment with a single cell which (due to the initial stored information) had reached a state when the mitotic operon and two functional operons (1 and 2) were active. This state was determined as follows:

$$C_0 = C_1 = C_2 = X_0 = X_1 = X_2 = P_0 = P_2$$
$$= R_0 = R_1 = R_2 = Y_0 = Y_1 = 1, \quad Y_2 = 0.5.$$

All other variables were zero at the starting time.

Three variants of this group will be presented in which the duration of the initial program was for 5, 105 and 205 steps of the computer.

The results obtained may be presented as intersections of the original cylinder (Fig. 2). The most important result in all three experiments is the

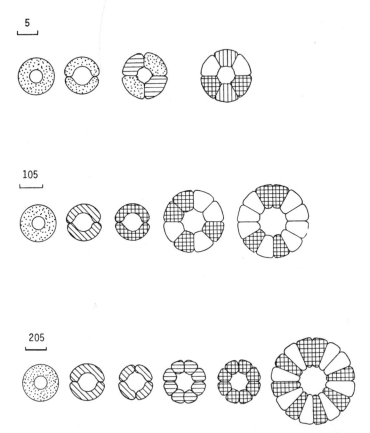

Fig. 2. Intersections of the cylindrical model showing the emergence of different cellular types in the development. The numbers at the beginning of the rows indicate the duration of the "stored" information in the computer steps.

emergence of different cellular types (differentiated cells) after several ce divisions. As may be seen, the number of divisions before "cytodifferentiation occurs depends on the time for which the initially stored information maintained. The longer this period, the greater the number of cellular division which will produce equipotential cells, and the greater the size of the cellula system thus obtained. The terminal state of the operons depended also on th duration of the initial program. In the third variant, only two different cellular types were obtained while in the first two variants three cellular types appeared. In this experimental group a steady state was reached in which the sixth functional operon was not yet deblocked.

3.2. Second group

The genetic nets were represented by the following matrices:

$$[\sigma_{ij}] = \begin{bmatrix} 0 & 0.20 & 0.20 & 0 & 0.20 & 0 & 0.20 & 0.10 \\ 0.20 & 0 & 0 & 0 & 0 & 0 & 0 & 0 \\ 0 & 0 & 0 & 0.20 & 0 & 0 & 0 & 0 \\ 0 & 0 & 0 & 0 & 0 & 0 & 0 & 0 \\ 0.20 & 0.20 & 0 & 0.20 & 0 & 0.20 & 0 & 0 \\ 0 & 0 & 0 & 0 & 0 & 0 & 0 & 0 \\ 0.20 & 0.20 & 0 & 0 & 0.20 & 0.20 & 0 & 0 \\ 0 & 0 & 0 & 0 & 0 & 0 & 0 & 0 \end{bmatrix},$$

$$[\kappa_{ij}] = \begin{bmatrix} 0.99 & 0 & 0 & 0 & 0 & 0 & 0 & 0 \\ 0.99 & 0.99 & 0 & 0 & 0 & 0 & 0 & 0 \\ 0 & 0 & 0 & 0.99 & 0 & 0 & 0 & 0 \\ 0 & 0 & 0 & 0 & 0.99 & 0 & 0 & 0 \\ 0.99 & 0 & 0 & 0 & 0.99 & 0.99 & 0 & 0 \\ 0 & 0 & 0 & 0 & 0 & 0 & 0.99 & 0 \\ 0.99 & 0 & 0 & 0 & 0.99 & 0 & 0.99 & 0 \\ 0 & 0 & 0 & 0 & 0 & 0 & 0 & 0 \end{bmatrix}.$$

The initial state was determined as follows:

$$C_0 = C_1 = C_2 = X_0 = X_1 = X_2 = P_0 = P_1 = P_2$$
$$= R_0 = R_1 = R_2 = Y_0 = Y_1 = Y_2 = 1.00,$$

and all other variables were zero.

In the first variant of this group, the influence of the initial number of cells on the terminal differentiation was studied. The threshold values of A_j for the different operons were as follows:

j	0	1	2	3	4	5	6	7
A_j	0.2500	0.2400	0.7200	0.0005	0.3600	0.0005	0.7200	1.000

Different results were obtained if the initial cellular group was a single cell, two or three cells. In the first case, a stable state was not yet reached even at the 1412th step (most of the cells had their mitotic operons still derepressed).

Starting with two or more cells, a stable state was soon reached at the 1175th step. The final number of cells, the number of cellular types and the cellular configuration obtained thus depended on the initial number of cells. This may hold true, however, only when the initial number of cells is small as in our experiments.

In other variants of this group, the effect of changing the threshold values A_j was studied. Only A_j of the mitotic operon (0) and of two functional operons (1 and 4) was changed. The results show that the program of development is very sensitive to extremely small variations in the threshold values A_j of the repressors. Even variations of these critical values less than 2% can drastically change both the final state of differentiation and the cellular configuration obtained.

In terms of phenotypic expression, these results indicate that very small variations in the parameter A_j can drastically change the size, the cellular configuration and the function (operon pattern) of a cellular system.

3.3. *Third group*

In this experiment the effect of cellular contact was studied. In the initial state, two different types of cells were considered, each of which had reached a steady state without having been in contact with the other. Only at the starting time of the experiment were they brought into contact. The matrices $[\sigma_{ij}]$ and $[\kappa_{ij}]$ were as follows:

$$[\sigma_{ij}] = \begin{bmatrix} 0 & 0.05 & 0 & 0.05 & 0.05 & 0 & 0.05 & 0.01 \\ 0.05 & 0 & 0.05 & 0.05 & 0 & 0 & 0 & 0 \\ 0 & 0 & 0 & 0 & 0 & 0 & 0 & 0 \\ 0.05 & 0.05 & 0.05 & 0 & 0 & 0 & 0 & 0 \\ 0.05 & 0 & 0 & 0 & 0 & 0.05 & 0.05 & 0 \\ 0 & 0 & 0 & 0 & 0 & 0 & 0 & 0 \\ 0.05 & 0 & 0 & 0 & 0.05 & 0.05 & 0 & 0 \\ 0 & 0 & 0 & 0 & 0 & 0 & 0 & 0 \end{bmatrix},$$

$$[\kappa_{ij}] = \begin{bmatrix} 0.99 & 0 & 0 & 0 & 0 & 0 & 0 & 0 \\ 0.99 & 0 & 0.99 & 0 & 0 & 0 & 0 & 0 \\ 0 & 0 & 0 & 0.99 & 0 & 0 & 0 & 0 \\ 0.99 & 0 & 0 & 0.99 & 0 & 0 & 0 & 0 \\ 0.99 & 0 & 0 & 0 & 0 & 0.99 & 0 & 0 \\ 0.99 & 0 & 0 & 0 & 0 & 0 & 0.99 & 0 \\ 0 & 0 & 0 & 0 & 0 & 0 & 0.99 & 0 \\ 0 & 0 & 0 & 0 & 0 & 0 & 0 & 0 \end{bmatrix}.$$

The following threshold values A_j were used:

j	0	1	2	3	4	5	6	7
A_j	0.1980	0.3900	0.0005	0.4000	0.3900	0.0000	0.4000	1.0000

The two cells were in a steady state, i.e., with the mitotic operon inactive (deblocked but repressed) and one functional operon active (deblocked and derepressed): no. 1 in one of the cells and no. 4 in the other. After the two cells came into contact, the steady state was disturbed (due to diffusion of metabolites between the two cells), and cellular divisions started leading to a new group of cells with new cellular types (Fig. 3).

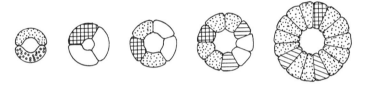

Fig. 3. Effect of the contact between cells.

All these computer experiments prove clearly that the molecular mechanism involved in our model can indeed account for the basic phenomenon characterizing cytodifferentiation; namely, the emergence of new cellular types through successive cellular divisions without changes in the informational content of DNA. At the same time, the essential features of this process are predicted by such a model. Our model as a whole has also some important implications for several other biological phenomena which seem to be intimately connected with the mechanisms underlying cellular differentiation [see 6, 7].

References

[1] Bl. Sendov and R. Tsanev, Computer simulation of the regenerative processes in the liver, *J. Theoret. Biol.* 18 (1968) 90–104.

[2] Bl. Sendov, R. Tsanev and E. Mateeva, A mathematical model of the regulation of cellular proliferation in epidermis, *Bull. Inst. Math. (Sofia) Bulg. Acad. Sci.* 2 (1970) 221–224 (in Bulgarian).

[3] R. Tsanev and Bl. Sendov, A model of the regulatory mechanism of cellular multiplication, *J. Theoret. Biol.* 12 (1966) 327–341.

[4] R. Tsanev and Bl. Sendov, Computer studies of the mechanisms controlling cellular proliferation, in: *Effects of Radiation on Cellular Proliferation and Differentiation* (Internl. At. Energy Agency, Vienna, 1968) 453–461.

[5] R. Tsanev and Bl. Sendov, A model of cancer studied by a computer, *J. Theoret. Biol.* 23 (1969) 124–134.
[6] R. Tsanev and Bl. Sendov, Possible molecular mechanism for cell differentiation in multicellular organisms, *J. Theoret. Biol.* 30 (1971) 337–393.
[7] R. Tsanev and Bl. Sendov, An epigenetic mechanism for carcinogenesis, *Z. Krebsforsch.* 76 (1971) 299–319.

A MATHEMATICAL SCHEME FOR MORPHOGENESIS: STRUCTURAL STABILITY AND CATASTROPHES

R. THOM

Institut des Hautes Etudes Scientifiques, Bures-sur-Yvette, France

The mathematical scheme proposed in order to explain the biological morphogenesis follows the tradition of those biologists who, from Aristotle to Goethe, saw in the living forms the realization of the universal archetypes. Actually, one endeavours to explain these forms by the constraint of structural stability, which can be expressed mathematically. The authors who have recently expressed analogical ideas are d'Arcy Thompson [1], Waddington [4] (who created the notion "chréode"), and Delbrück [2] (who suggested an explanation for cellular differentiation by a stable regime of metabolism).

Let U be the domain of space–time including the studied morphological processes, and let u be a point from U. We suppose that the set of local processes (the *germs* of the processes) around a point such that $u \in U$ can be parametrized by the points of a smooth manifold F (supposed compact and finite dimensional).

The local evolution of these dynamic germs is described by a field of vectors X in F. The totality of the possible local evolutions – compatible with the external conditions on the boundary of U – is described by a continuous family of dynamics $X(x; u)$ on $U \times F$, i.e., a continuous mapping G from U into the space $D(F)$ of dynamics over F. A global evolution over U is defined when a section $S: U \to U \times F$, continuous in general, is given. Actually, the fact that the local state of the process is a state of local balance is expressed as follows: the trajectory of the flow X for the point $s(u)$ is considered, and its asymptotic state for $t = +\infty$ is studied. In many cases (the most interesting ones), the asymptotic state is defined by an attractor A of the dynamic X (for example, an attracting limit point, or an attracting closed trajectory). It may happen that this attractor A is structurally stable: for every field X' close enough to X (in the C^1 topology), there exists an attractor A' neighbouring A with a homeomorphism h from a neighbourhood of A onto a neighbourhood of A' which maps a trajectory of X on a trajectory of X'. If at the point $u \in U$ the asymptotic state $\bar{s}$ of the section $s(u)$ is given by a structurally stable attractor, the same holds for every point u' close enough to u if s is continuous. In this case, u is called a *regular point* of the morphology. The regular points form an open set in U; its complement is called the set of the *points of catastrophe*. If v is a point of catastrophe, either the section $\bar{s}(v)$ defines an attractor not structurally stable (a *point of bifurcation*) or the section $\bar{s}(v)$ is discontinuous in v (a *point of conflict*).

Every empiric morphology is characterized by a set of points, at which the medium has qualitative discontinuities; one endeavours to identify these points with the sets of catastrophe by the mathematical theory. For this reason, we need some hypotheses for the nature of the local dynamics X, for the mapping G from U into $D(F)$, and for the choice of the dominant attractor $\bar{s}(u)$, which can define the set of conflict. A complete mathematical theory exists only if the local dynamics X are dynamics of gradients, if the mapping G is in general position with respect to the set of bifurcation of the dynamics, and if the set of conflict is defined by a simple rule like the *rule of Maxwell*: the dominant attractor in a point $u \in U$ is the minimum of a potential on F, which has the lowest value. So the notion of universal unfolding of a singularity of a function (which is notably explained by Mather [3]) permits us to specify the topological nature of the sets of catastrophe. When the dimension of U is less than 4, there are 7 types of singularity which can be represented generically as catastrophes of bifurcation. These are the 7 elementary catastrophes which I have enumerated. One now knows these singularities for dim $U \leqslant 15$.

The application of the theory to the empiric morphologies (especially to material morphologies, living or not living) raises numerous problems that demand a generalization of the preceding scheme. A fundamental difficulty arises in connection with the notion "transition of phase". A "local phase" is characterized by a pseudo-group of invariance. Therefore, the groups of local isomorphisms of the pseudo-group take effect in the variety F of internal states and it is possible to create a theory of bifurcation of invariant functions by means of symmetries, as well as of "breaks" corresponding to symmetries. These effects have already occurred in the transition liquid–gas since the elementary "umbilic elliptic" singularity does not occur but it is replaced by an umbilic parabolic section which has a symmetry of revolution (breaking of a jet). In biology, the embryological development can be described in the beginning only through elementary catastrophes. Later on, catastrophes with symmetries (the first is the bilateral symmetry of the embryo), involving namely the metric group or its sub-groups (morphogenesis of the bone, of the articulations, of the eye), take place.

References

[1] d'Arcy Thompson, *On Growth and Form* (Cambridge Univ. Press, London, 1945).
[2] M. Delbrück, Unités biologiques donées de continuité génétique, Colloque CNRS, Paris (1949).
[3] J. Mather, Notes on topological stability, Harvard University, Cambridge, Mass. (1973).
[4] C. H. Waddington, *Strategy of the Genes* (Allen and Unwin, London, 1957).

MODELS OF THE DEPENDENCE OF HOSPITAL UTILIZATION UPON MEDICO-SOCIAL FACTORS

I. VÄÄNÄNEN, G. BÄCKMAN, A. S. HÄRÖ, J. PERÄLÄ and O. VAUHKONEN

The Finnish Hospital League, Helsinki, Finland

1. Introduction

In many countries, continuous rises in the capacity of hospitals and in the expenditure entailed have been common phenomena during the last twenty years. As these increases have generally occurred more rapidly than the rise in the gross national product, it is understandable that public discussion has been aroused in regard to the role and the function to hospitals in health systems.

One of the most fundamental problems relates to the extent of hospital services required properly to satisfy the needs of a given population. The authors of this study are striving to find a method for the definition and measurement of the factors that influence the need of hospital services. The strategy of the approach to the problem is based upon the observation that variations exist in the utilization of hospital services in different regions, even if the facilities available are almost the same [4].

With this observation as a starting point, the authors have aimed at the construction of an ecological level model for the explanation of the regional differences in the utilization of hospital services, based upon quantifiable and predictable variables. The second objective of the study is that of predicting the future regional hospital bed requirements by means of the model so constructed; a condition for the fulfilment of this aim is the existence of a regionalized hospital system in which knowledge exists of the population using the hospital. The Finnish hospital system offers this advantage to researchers.

2. The Finnish hospital system

In Finland, the commune (local authority) is the smallest administrative entity legally responsible for the provision of health services to the population.

The communes are combined to form 21 hospital regions, each of which owns and maintains a central hospital. The central hospitals are general hospitals which function on a specialist level. In addition to these, almost all communes have their own general hospitals of smaller type.

In Finland, hospitals are financed from taxation. The communes are supported by the state to the extent of a fixed percentage of the capital and running costs of hospitals. These subsidies are determined on the basis of the financial resources of the communes, and they vary between 39 and 70%. The patient pays a fixed sum of about $2.50 per day, which covers all possible services, and is the same in all hospitals. This daily payment corresponds to about the legal minimum wages for two hours' work, and even this is defrayed by the welfare authorities if the patient is without means. Consequently, the direct personal costs of hospital services do not appear as a noticeable obstacle to hospital admission in Finland, and have not been taken into account in this study.

3. The conceptual background

When the aim is to develop a mathematical expression, an explanatory model, for the utilization of hospital services, the first phase is to make a general examination of the hospital function with a view to the determination of the type of model that could come into question. In a previous study [4], the authors showed that within a regionalized hospital system the demand for hospital care is satisfied in an order such that the most urgent needs are satisfied first (Fig. 1). Consequently, the growth of the hospital capacity has systematic consequences in the selection of patients (Fig. 2). As hospital capacity grows, several patient groups reach the saturation level in definable order, whereas for some patient groups the demand does not show any signs of balance. These statements constitute the methodological point of departure for this study.

The unit of measurement chosen for hospital utilization was the number of admissions to a hospital (equivalent to hospital visits and discharges) per thousand of the population. An alternative would have been to measure the hospital utilization by hospital days (patient days). However, the number of hospital days is in this connection extremely sensitive to secondary influences, whereas admission to the hospital usually occurs at the acute phase of illness, when the medical indications are most pronounced. Consequently, it was considered that the number of admissions is related to the need for hospital services to a greater extent than the number of hospital days. The number of hospital beds needed is to be derived from the number of admissions by means of the average length of stay.

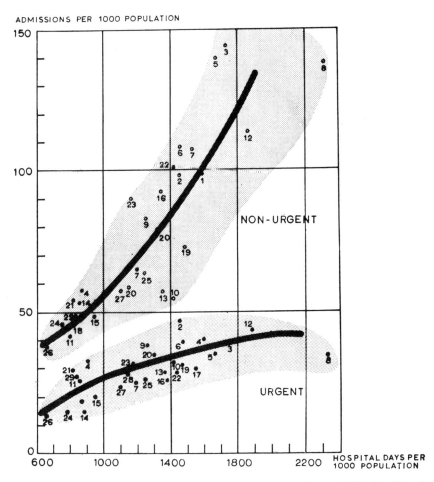

Fig. 1. Division of patients in urgency groups by the level of total hospital utilization (Kuopio Central Hospital District, 1960).

The population of the commune was chosen as the unit on the ecological level. The whole material was divided into groups according to the patients' commune of domicile, and the extent of hospital utilization in a region was ascertained by combining the results obtained from the communes concerned.

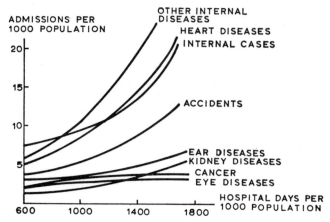

Fig. 2. Division of patients in diagnosis groups by the level of total hospital utilization (Kuopio Central Hospital District, 1960).

4. The material

This study is solely concerned with the use of general hospitals. Patients admitted to tuberculosis sanatoria or mental hospitals have not been included.

The authors have examined the data with regard to the hospital visits in all the communes within every hospital region in Finland in 1960 and in 1967. The total number of communes is 546. The material comprises the routine discharge reports of hospitals as coilected by the National Board of Health. The number of patients treated in the general hospitals was 468 000 in 1960, and 560 000 in 1967.

5. The Methods

5.1. *A theoretical approach*

In an attempt to develop mathematical models which explain hospital utilization, the first phase is to examine the possibilities of using single equation models. Feldstein [2] has presented models of this type, with a difference in composition. The authors of this study have also developed single equation regression models, both of linear and of saturation type [5]. However, efforts to solve the problem by applying models of this type proved unsatisfactory. In fact, this result is in agreement with the statements presented in the conceptual background of this study. It was stated there that with growing hospital capacity, several patient groups get their demand for hospital services saturated in a definite order, which depends upon the urgency of needs. Thus, for the explanation of hospital utilization, multiple equation recursive regression models are to be constructed.

Since the same resources are used for the care of patients of different preference groups, the admission of urgent cases limits the possibilities of admitting non-urgent patients. As a result, the regression models that explain the hospital use of various patient groups cannot be mutually independent. The multiple equation recursive model takes these intercorrelations into account. In such a model, various components of the total consumption of hospital services will be estimated at various stages, in such a way that the regression equations form a causal chain. This recursive approach can be presented by the following equations:

$$y_{1t} = a_{10} + L_{1t} + e_{1t},$$
$$y_{2t} = a_{20} + a_{21} y_{1t} + L_{2t} + e_{2t}, \qquad (1)$$
$$\vdots$$
$$y_{mt} = a_{m0} + a_{m1} y_{1t} + \ldots + a_{m, m-1} y_{m-1, t} + L_{mt} + e_{mt},$$

in which

$y_{1t}, y_{2t}, \ldots, y_{mt}$ = the dependent variables (hospital bed requirements of various groups during a period of time t),

$L_{1t}, L_{2t}, \ldots, L_{mt}$ = the functions of explanatory variables during period t,

$e_{1t}, e_{2t}, \ldots, e_{mt}$ = error factors during period t,

$a_{10}, a_{20}, a_{21}, \ldots, a_{m, m-1}$ = numerical quantities (parameters).

5.2. *Dependent variables*

In the pursuance of the theoretical approach, the hospital utilization, that is, the variable to be explained, had to be divided into parts applicable in a multiple equation recursive model. According to the statements in the conceptual background, the logic in the causal chain of the models should be related to the urgency of the patient groups. Consequently, all the hospital visits to be studied were divided into three preference groups, in conformity with the urgency revealed by diagnosis: urgent, non-urgent and chronic cases. Groups of this kind are not mutually exclusive, but form clusters with differences of sufficient significance to be of value.

In this study, the group of *urgent cases* relates to emergency, delivery and terminal cases. *Emergency cases* include codes 420.1, 420.2, 433, 434.2, 540.1,

Table 1

ICD 7th Rev.	ICD 8th Rev.	Age at least	ICD 7th Rev.	ICD 8th Rev.	Age at least
A 4	010–012		A 80	394–398	
A 5	017–018	50	A 81	410–414	50
A 6–10	090		A 82	420–429	50
A 44	140–149		A 83	400, 402	
A 45	150			404	50
A 46	151		A 84	401, 403	50
A 47	152–153		A 85	440–448	50
A 48	154		A 93	490–491	50
A 49	161		X 23	515–516	
A 50	162		X 24	518	50
A 51	174		X 25	519	
A 52–53	180–182		A 99–100	531–532	50
A 54	185		A 105	571	
A 56	170–171		A 109	581–584	
A 57	190–199		A 110	590	50
	155–160		A 112	600	65
	163, 172–173		A 114	591	
	183–184		X 28	694, 707	
	186–189		A 112	712–714	
A 58	204–207		A 123	717–718	
A 59	200–203			728	
X 11	507, 692.9		A 125	735	
	493	50	X 29	722	50
A 63	250			725–726	
X 12	243–246			330, 733	
	251–258		A 127	741	
X 13	208, 286–289		A 128	746–747	
X 20	301, 304	20	A 129	750–759	
	308, 306		A 136	794	
A 70	430–438		A 137	780–793	65
A 72	340		A 138	800–804	
A 73	345		A 139	805–809	
A 76	375		A 140	820–824	50
X 22	321–324		A 150	950, 958	
	342–344				
	331–333				
	347				
	353–354				

541.1, 550, 561, 570, 576, 587.0, 650–652 and 800–999 of the WHO's International Classification of Diseases (ICD) [7]. *Deliveries* include codes 660 and 670–678. *Terminal cases* are defined as patients who die in hospital from causes other than the urgent ones previously mentioned. This information was also readily available.

In principle, the group of *chronic cases* was formed in the same way, with the coded diagnoses and the age of the patient constituting decisive factors. The group of chronic cases consists of (a) patients with more than 30 days' length of stay in the hospital, and (b) patients with diagnoses belonging to the groups of the ICD shown in Table 1.

The *non-urgent* group was calculated by subtracting the groups of urgent and chronic cases from the total.

By means of this, the total hospital utilization was divided into three dependent variables:
  urgent cases (URG),
  non-urgent cases (NONU),
  chronic cases (CHRO).
The first of these variables (URG) represents the highest degree of urgency, and the last (CHRO) the lowest. Each of these three variables will be explained by a regression equation; in combination they will form the recursive model for the explanation of total hospital utilization.

5.3. *Explanatory variables*

The selection of the variables determining the demand for hospital care of the patient groups corresponding to the three dependent variables was founded upon the analysis of the communal variations in the consumption of hospital services. The variables were empirically tested by regression analysis during the construction of the model, and only those of them which displayed a significant correlation with the dependent variables were selected for incorporation into the final model.

The following list describes the variables. Each of them refers to the population of a commune and the numerical values are the rates per 1000 of the population.

(a) *Population* (POP).

(b) The concept of *availability of hospital services* was measured by means of the variable describing the number of general hospital beds "owned" by the commune (SUPPLY).

(c) The concept of *availability of ambulatory services* was not directly measurable. National Health Insurance provides for doctors' fees, and the number of persons reimbursed during a calendar year can be obtained (AMB).

(d) The *illness* concept is a difficult one, and can be studied in terms of the medical, psychological and sociological concepts of illness. It has been shown [3] that correlations among the factors extracted with a view to the representation of these different concepts of illness indicate a strong mutual dependence of different manifestations of illness. In this study, the factor was represented by a measurement of the number of persons who had obtained daily allowances for sickness from the National Health Insurance during a calendar year (SICK).

(e) *Educational* concept. It is reasonable to assume that a higher level of education is positively correlated with the perceived illness and the perceived value of the treatment. This factor was measured by the variable EDU, which gives the number of persons graduated from intermediate school (corresponds to nine years of school).

(f) The concept of *health consciousness*. In previous studies [6], it has been demonstrated that hospital use as a consequence of childbirth and illnesses related to pregnancy is correlated with the hospital use of the whole population. The consumption of maternal and child health services seems to increase the health consciousness of the whole family [1]. The variable DEL gives the number of hospital visits as a result of childbirth and illnesses related to pregnancy, and is used in this study as a measurement of the health consciousness (or level of health education).

(g) The concept of *old age*. It is known that older persons require more hospital services than do younger age classes. The variable which indicates the contribution of persons of 65 years and over (OLD) is a natural component in the model. However, the figures in themselves can be misleading, since the large relative contribution of old persons may reflect a lower mortality and a better health situation during the earlier part of life.

(h) The concept of the possibilities of having *other types of care*. In the higher age classes particularly, cases occur which cannot be treated at home, but which do not in fact require hospitalization. The selected variable was the number of beds for sick persons in homes for the aged in communes (OBED).

5.4. *The construction of empirical models*

The variables described above were measured (if applicable) for the years 1960 and 1967. There were 546 communes to be taken into account. The empirical measurements were applied to the theoretical model separately for the years 1960 and 1967.

In the formulation of the empirical models, selective regression analyses were used. The parameters were estimated by the least squares method, with one predetermined parameter being employed at a time in the model, in a fixed order. The significance of the estimated coefficient for the new variable was subjected to the F-test. When a new variable was added to the model, a test was made of whether the variables added earlier still had a significant effect. When conclusions were drawn, a probability of 0.05 was used as the risk level. In the study, the variables were generally preserved in their original scale, without transformation into logarithms or other types of presentation.

The following empirical recursive multiple equation models were estimated:

For the year 1960:

$$URG = 10.554 + 0.103\ SUPPLY + 0.278\ POP + 0.133\ DEL$$
$$- 0.108\ OLD, \qquad R = 0.639, \qquad (2a)$$
$$NONU = 7.830 + 1.322\ URG + 0.581\ SUPPLY + 0.192\ OBED$$
$$+ 0.045\ OLD - 0.027\ DEL, \qquad R = 0.701, \qquad (2b)$$
$$CHRO = 3.092 + 0.387\ NONU - 0.053\ URG + 0.024\ DEL$$
$$- 0.001\ EDU + 0.0007\ OBED, \qquad R = 0.896. \qquad (2c)$$

For the year 1967:

$$URG = 6.392 + 0.128\ SUPPLY + 0.038\ POP - 0.067\ OLD$$
$$+ 0.076\ DEL + 0.045\ AMB + 0.005\ SICK,$$
$$R = 0.483, \qquad (3a)$$
$$NONU = -2.947 + 0.673\ URG + 0.456\ SUPPLY + 0.259\ AMB$$
$$+ 0.015\ SICK + 0.164\ DEL, \qquad R = 0.613, \qquad (3b)$$
$$CHRO = -9.522 + 0.541\ NONU + 0.487\ URG + 0.124\ OLD$$
$$+ 0.028\ POP - 0.060\ DEL + 0.009\ SICK,$$
$$R = 0.937. \qquad (3c)$$

Fig. 3. illustrates the function of the models.

In the estimation of the two empirical models, consideration was given to the practicability of comparing the permanency of the explanatory significance

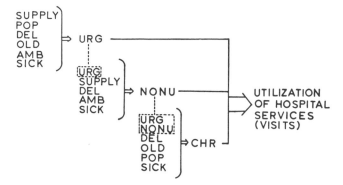

Fig. 3. Scheme illustrating the function of the empirical recursive models.

of different variables. It can be assumed that the variables which are significant both in 1960 and 1967 will also be of importance in years to come.

In the explanation of the hospital use of *urgent cases* (eqs. (2a), (3a)), the variables SUPPLY and OLD are of significance. The former has a positive sign, but the latter a negative one; this indicates that these variables influence the hospital use of urgent cases in opposite directions. The variables DEL and POP possess explanatory significance in both groups of material.

In the model for *non-urgent cases* (eqs. (2b), (3b)), the variables URG and SUPPLY possess a marked and permanent explanatory significance. However, in 1960 the variable DEL has a negative, and in 1967 a positive sign; this indicates that the explanatory significance of this variable is uncertain, at least as far as this material is concerned.

In the model for *chronic cases* (eqs. (2c), (3c)), the variable NONU has a strongly positive correlation both in 1960 and 1967. In 1960 the variable URG has a negative sign, and in 1967 a positive one. This may reflect the differences in the availability of hospital services in 1960 and 1967. In 1960, the number of beds was considerably less, and the hospitalization of urgent cases limited the possibility of hospitals to accept chronic cases. In 1967, these interrelationships were less influential by reason of the increased resources available.

6. The use of the recursive model in regional comparisons and forecasts

The models can be employed in numerous ways in practical planning but our attention has been mainly directed along the problem of prediction. The explanatory variables have been estimated for the year 1980, with the parameters based upon the situation in 1967.

In practical applications of the recursive model, only two sub-models were used. The third group, that of chronic cases, was incorporated in the non-urgent group. The model for 1967 was employed, as it covered more of the logically relevant aspects (AMB and SICK). The variable DEL was excluded from the model for non-urgent cases, by reason of its uncertain significance in this group. Correspondingly, the constant term was adjusted. The final versions of the empirical equations used for the prediction of future hospital utilization were as follows:

$$URG = 6.382 + 0.128 \text{ ESTB} + 0.038 \text{ POP} - 0.067 \text{ OLD}$$
$$+ 0.076 \text{ DEL} + 0.045 \text{ AMB} + 0.005 \text{ SICK}, \qquad (4a)$$
$$NONU = 17.094 + 0.673 \text{ URG} + 0.455 \text{ ESTB} + 0.259 \text{ AMB}$$
$$+ 0.015 \text{ SICK}. \qquad (4b)$$

In these equations, the explaining variables contain a new one: the estimated amount of beds (ESTB). This seemingly peculiar arrangement results from the model predicting the visits and not the beds; moreover, the factor of available resources (SUPPLY) possesses great explanatory significance. Consequently, an approximate estimate of beds at the relevant moment in the future is required. For this purpose, an extrapolation was used of the regional trends between 1960 and 1967. The model thus corrects the usual trend extrapolations on the foundation of facts with regard to the population concerned.

The numerical values for the explanatory variables for the future were compiled from different sources. The population estimates (POP, OLD) were available (State Planning Office), and the estimated number of deliveries for the calculation of the variable DEL has been published by the State Statistical Centre. The variables AMB and SICK were estimated from the yearly mean growth (2.5%) of the relevant health insurance data measured between 1965 and 1971.

For the year 1980, the model gives the expected numbers of hospital visits for any given population. It predicts the number of cases of sickness among the population that will need hospital care, but not the actual number of beds. For this reason, the numbers of visits need to be transformed into hospital days, and ultimately into beds. This can be effected by multiplying the number of visits by the mean length of stay, and dividing the number of patient days so derived by 365, or preferably by 330, which gives a direct estimate of the required number of beds on the assumption that the occupation rate is 90%.

In the transformation of hospital visits into the corresponding number of beds, the mean length of stay has to be estimated for the relevant date in future. Scientific development results in an improvement of the methods of examination and treatment, which can be expected to shorten the mean length of stay. Nevertheless, more effective treatment keeps the very severe cases alive and prolongs their hospital treatment time. Consequently, there is hardly reason to expect that the mean length of stay of acute cases will be markedly changed in the near future. The problem is more complicated with respect to chronic, long-stay cases. Each of these two contradictory influences will have even more accentuated consequences in the treatment of chronic cases. It is difficult to predict how these effects will be balanced in days to come.

Notwithstanding these restrictions, the model increases the reliability of estimations of future bed requirements. The influence of numerous variables is measurable, and only the length of stay variable is estimated separately. It may even be of advantage to possess an opportunity to manipulate the length of stay variable for further investigations, e.g. by simulation techniques.

We have used three alternative predictions for the mean length of stay:

(i) The mean length of stay of the patients of each commune will not evince any change between 1967 and 1980.

(ii) The change in the national mean length of stay was measured between 1960 and 1967. This trend was extrapolated until 1980, and the mean length of stay of the patients in each commune was adjusted accordingly.

(iii) The national mean length of stay, calculated for the year 1980, was applied to every commune.

The model has been empirically applied in 12 hospital regions in Finland. The main interest has been in the future estimations, with 1980 being selected as the target year.

By means of the data compiled separately for each commune, the model has given the number of yearly hospital visits per 1000 of the population for each commune in 1980. The figures obtained have been multiplied by the mean lengths of stay estimates (a, b, c), and by the size of the population. The number of hospital days so derived has been transformed into beds, on the assumption of 90% occupancy rate. The number of hospital beds for a region was made up by the figures relative to the communes that form the region.

Table 2

Number of hospital visits calculated with a recursive multiple regression model, the number of hospital beds required, and the population in 1980 in 12 hospital regions in Finland. The corresponding situation in 1967 is also given.

	1980					1967		
Region	Hospital visits	Hospital beds a	b	c	Population	Population	Hospital visits	Hospital beds
Helsinki	191	7.9	8.8	8.9	1201310	967374	136	5.4
Turku	142	5.4	6.0	6.6	416533	409752	118	4.3
Tampere	150	6.4	7.1	7.0	419719	393644	138	5.5
Hämeenlinna	141	5.2	5.8	6.7	157774	156907	121	3.8
Satakunta	142	4.5	5.0	6.6	253500	241519	126	3.6
Keski-Suomi	155	5.0	5.5	7.2	260162	248066	136	4.6
Seinäjoki	143	4.3	4.7	6.7	173899	205832	121	3.5
Kainuu	126	4.1	4.6	6.0	105447	108400	116	2.7
Lappi	156	5.8	6.2	7.2	158800	132342	153	5.2

Table 2 lists the calculated hospital bed requirements for 12 hospital regions in 1980. Eqs. (4a) and (4b) were used, and the parameters relate t circumstances in the year 1967. The table also indicates the population in 1967 and 1980, and the current hospital beds per 1000 of the population in each region.

The estimated numbers of beds are markedly lower than those which could be calculated with the extrapolated national trends. Some of these results have been applied to practical decisions concerning future hospital capacity, and some districts are using the model, corrected in accordance with the latest data (1970), for their long-term plans.

7. Concluding remarks

It seems practicable to construct multiple regression models on an ecological level which are to a reasonable extent serviceable in the estimation of hospital bed requirements. The approach made here is particularly aimed at a model which takes into account the preferences for admission with respect to the various groups of hospital users. The multiple equation recursive model, which first explains hospital visits for urgent causes, and with the result taken into consideration explains the non-urgent visits, gives meaningful figures on empirical applications. On the average, an explanation is derived for about 80% of the hospital visits of a defined population. As sociological models concerned with people's behaviour seldom attain higher explanatory capacity, the model presented here could be regarded as reasonably suitable for use.

The successive saturation of the needs of different preference groups of hospital users is rather obvious in regionalized hospital systems such as that in Finland. In a system with no distributional responsibilities, it might be a less obvious phenomenon, but it must quite certainly exist to some extent in systems organized along appreciably different lines. For example, it could be detected in an analysis of the preference scale applied by physicians who use a hospital for their private patients. The cost factor can be of great importance in differently organized systems, and some other specific aspects of each health service system should certainly be considered in this connection. In this study, the mean length of stay was estimated without any special mathematical model. Such models can be constructed, and could probably improve the accuracy of the predictions.

This study is not intended to provide hospital researchers in differently organized health service systems with a ready-made model. The model is not a final one, even for our own purposes, and others prepared with different components need to be studied. The principal aim has been that of demonstrating that analyses of hospital use can provide useful information with regard to the difficult question: How many beds are needed? This type of model may enable researchers to provide those making the decisions with conclusions of more use than the pessimistic statement: more beds means more needs.

The model presented here is concerned with the total consumption of hospital services. Consequently, it does not distinguish between the figures which should indicate the bed requirements for narrow specialities. These figures are incorporated in the results obtained by the model, and it merely becomes a question of medical policy to decide how they are to be separated and formed into special wards.

As the predictions given by the model are based upon the characteristics of the population in question, it could be expected that they will be of greater value in regional health planning than the general norms or standards. The utilization of the model would thus facilitate the defining of the objectives in regional hospital planning. The model approach supports those decision makers who are endeavouring to realize the modern concept of hospital administration with a definite aim in view.

References

[1] G. Bäckman, Men of working age as consumers of hospital services, Health Serv. Res. Natl. Board of Health in Finland (7) (1969) (in Finnish, with English summary).
[2] M. S. Feldstein, *Economic Analysis for Health Service Efficiency* (North-Holland, Amsterdam, 1967).
[3] E. Kalimo, Determinants of medical care utilization, *KELA* (Helsinki) A5 (1969) (in Finnish, with English summary).
[4] I. Väänänen, A. S. Härö, O. Vauhkonen and A. Mattila, The level of hospital utilization and the selection of patients in the Finnish regional hospital system, *Medical Care* 5 (1967) 279–293.
[5] I. Väänänen, G. Bäckman, S. Härö, J. Perälä and O. Vauhkonen, Explaining and predicting of hospital utilization by means of regression models, *Publ. Finnish Hosp. League* A2 (1969) (in Finnish, with English summary).
[6] O. Vauhkonen, Utilization of general hospital services by municipalities in 1960, Health Serv. Res. Natl. Board of Health in Finland (2) (1967) (in Finnish, with English summary).
[7] World Health Organization, *International Classification of Diseases*, 7th Rev.

A METHOD OF COMPUTER SIMULATION USED TO STUDY SOME SUBLETHAL EFFECTS OF IRRADIATION ON CELL KINETICS

A.-J. VALLERON

Unité de Recherches Statistiques, Institut National de la Santé et de la Recherche Médicale, et Laboratoire de Biostatistique de l'Université Paris 7, 94800 Villejuif, France

1. Introduction

1.1. *The cell cycle*

Only a few biological definitions are necessary to understand the models to be discussed below. The *cell cycle* begins when a cell is just born and ends when the cell is just going to divide into two new cells. The DNA content of the cell increases during the cell cycle (cells at the end of the cell cycle have twice as much DNA as those at the beginning). It is now well known that the DNA content increases only during a part of the cell cycle (synthetic phase S). Thus one usually distinguishes four phases during the cell cycle: the presynthetic phase (G_1 or gap 1), the synthetic phase (S), the post-synthetic phase (G_2) and the mitotic phase (M). This is summarized in Fig. 1. We can identify the

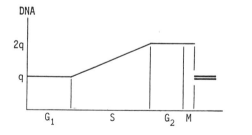
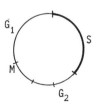

Fig. 1. The cell cycle. The DNA content increases during the S phase. G_1 – presynthesis; S – synthesis; G_2 – post-synthesis; M – mitosis.

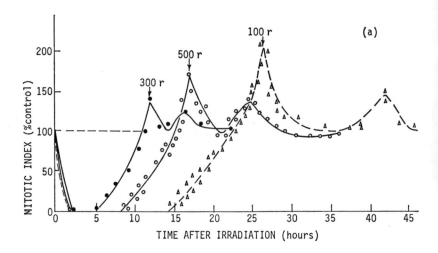
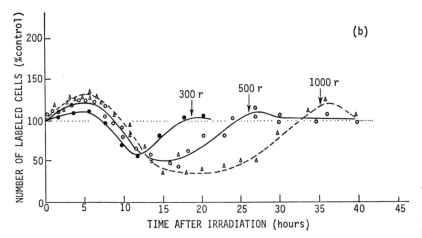

Fig. 2. Experimental data for mitotic and labelling indexes after 300, 500 and 1000 r irradiation. (From Kim and Evans [9].)

cells in the G_1, S, G_2 and M phases, respectively, by the following experimental methods:

(i) Morphologically it is possible to distinguish a cell in the mitotic phase from a cell in the intermitotic phase (G_1, S or G_2).

(ii) By using autoradiography, it is possible to distinguish a cell during the synthetic phase from other cells.

(iii) Cytospectrophotometry makes it possible to measure the DNA content of a cell; thus it is possible to distinguish between G_1, G_2 and S cells.

1.2. The mitotic delay after irradiation

After exposure to ionizing radiations, mitosis is delayed. This was first demonstrated by Canti and Spear [3] 45 years ago and, since then, by many authors on many different materials.

Fig. 2 shows the evolutions of the percentage of cells in the M phase after three different doses of acute irradiation; here one can observe an immediate decrease of the mitotic index followed by an overshoot of the mitotic activity. In general, the decrease of the mitotic index may or may not be immediate. This fact is interpreted as a consequence of the "block"; i.e., a part of the cells of the population has been "blocked" during a certain time under the effect of irradiation.

The exact mechanism of the mitotic delay is not yet known; it is probable that this mechanism is different for different cell lines. There have been many interpretations of this phenomenon, an exhaustive review of which has been given by Frindel and Tubiana [5].

The importance of setting up biomathematical methods to investigate the mechanism for a given cell line is obvious. From a theoretical point of view, a discovery, for example, of a part of the cell cycle where cells are blocked would allow us to study which unknown mechanism occurs in that precise part of the cell cycle. From a practical point of view, one can hope that these sublethal effects of irradiation can be used in order to synchronize, before the treatment, tumour cells in their most sensible phase to a given agent.

1.3. Principles of the computer method of simulation

The method and some examples of its application are described elsewhere [17]. Here we briefly state the principles: In Fig. 3, we represent the family tree originating from a single cell born at time $t = 0$; we assume first that there is neither a cell death nor a cell loss. Each branch of the tree represents a cell and is divided into four segments which correspond to each of the four phases of the cell cycle.

A two-dimensional array $Y(J,K)$ is defined by the following equations:

(i) $Y(J,K)$ = time necessary to reach the end of the Kth phase of cell J ($K \leq 4$),

(ii) the mother of the Jth cell is the J_1th cell, where $J_1 = [\frac{1}{2}J]$ (= the largest integer smaller than $\frac{1}{2}J$).

(iii) $Y(J,1) = Y(J_1,4) + \text{Ph}_1(J)$, $Y(J,K) = Y(J,K-1) + \text{Ph}_k(J)$ $(K \neq 1)$, where $\text{Ph}_k(J)$ is the duration of the Kth phase of the Jth cell.

The program
(a) Diagram of the progeny of cells:

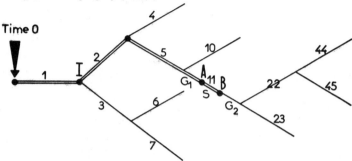

The mother of cell no. 45 is cell no. $[\frac{1}{2} \cdot 45] = [22.5] = 22$.

(b) Representation of the cell population in the computer:
$I = G_1$, $II = S$, $III = G_2$, $IV = M$.

Examples:

$Y(1, IV)$ = time elapsed between time 0 and time I (mitosis);
$Y(11, I)$ = time elapsed between time 0 and time A
   = $Y(5, IV) + G_1$ (end of the G_1 phase of cell no. 11);
$Y(11, II)$ = time elapsed between time 0 and time B
   = $Y(11, I) + S$ (end of the S phase of cell no. 11).

Fig. 3. The principle of identification of the cells of the progeny is shown in (a), the recurrence equations are described in (b).

(i) is a definition; (ii) is obvious if the identification numbers of the cells are chosen sequentially, generation after generation, as shown in Fig. 3. The array $Y(J, K)$ is completed by using the recurrence relations in (iii). The numbers $Ph_k(J)$ are pseudo-random numbers from suitable distributions.

The array $Y(J, K)$ is stored and may be used, for instance, to compute how many cells are present in each phase at time t. For instance, the cell J is present in S at time t if

$$Y(J, 2) < t < Y(J, 3).$$

It is possible to simulate cell loss or death in proportion p by defining a new array $D(J)$ as follows:

(iv) $D(J) = 1$ if cell J is dead or does not exist,
(v) $D(J) = 1$ if $D(J_1) = 1$, where $J_1 = [\frac{1}{2}J]$.

Condition (v) expresses that daughters of a dead cell do not exist. If we assume that a proportion p of the cells present at time t die, we let $D(J) = 1$ with probability p for the cell J present at time t.

This method makes it easy, for instance, to follow (within a family tree) the progeny of a particular cohort of cells. Various useful results for this family tree are stored. The whole procedure is repeated on successive family trees until the desired precision is obtained; one example will be given below in detail.

2. Model building

2.1. *Three models for mitotic delay*

We have set up three simple models of the mitotic delay which may be used either separately, or in any combination of them, appropriate to the basic ideas of the investigators who have interpreted their experimental data.

2.1.1. *Model (a): block*
Model (a) is an exact model of the blocking; in this method, it is assumed that there is a point B of the cell cycle where every cell present at the time of

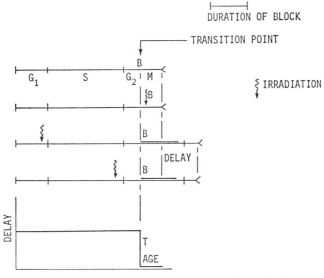

Fig. 4. Model (a): block model. The block occurs at the transition point between G_2 and M.

irradiation is blocked during a period T_B. This model is illustrated in Fig. 4; in this figure, for simplicity, the lengths of the cell phases and the duration of the block are assumed to be constant.

2.1.2. Model (b): repair

In Model (b), we assume that there is an important point R somewhere in the cell cycle. But cells, after irradiation, do not behave in the same way; when they are irradiated, we may assume for instance that an enzyme necessary for the transition in R is damaged. The cells irradiated at an age greater than R are not delayed and the cells irradiated before R are supposed to begin to

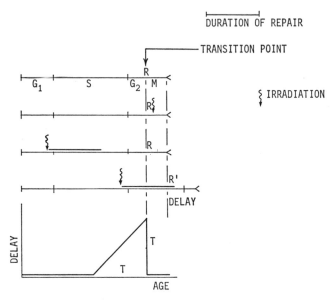

Fig. 5. Model (b): repair. The repair should be completed before the transition between G_2 and M.

repair the damage immediately; the time necessary to repair the damage is T_R. In this model, illustrated in Fig. 5, the cells irradiated at an age immediately before R are delayed more than the younger cells.

2.1.3. Model (c): lengthening of the DNA synthesis under irradiation

Sinclair [11], for instance, interprets part of his experimental data by assuming that there is a lengthening of the DNA synthesis under the effect of irradiation. In Model (c) we let v be a constant rate of increase of the DNA content of a cell. After irradiation the rate of increase changes to $v' = pv$, where p is a parameter of the model, as illustrated in Fig. 6.

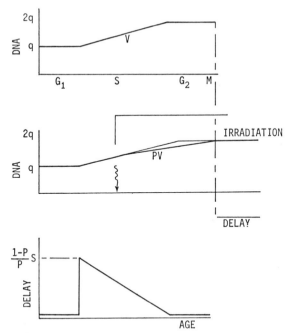

Fig. 6. Model (c): lengthening of DNA synthesis. After irradiation, the rate of synthesis is assumed to become $v' = pv$, where $p < 1$.

2.1.4. *Some remarks about Models (a), (b) and (c)*

We could think of various other models. However, when an investigator describes a "G_2 block", it appears that most of the time he refers to either Model (a) or (b). These two models have the advantage of including the usual references to block mechanisms, inhibition of the mitosis (B or R are the transitions between G_2 and M), inhibition of the synthesis (B or R are the transitions between G_1 and S). The third model assumes that the DNA synthesis may be lengthened by the radiation.

In the above three models, we have not taken into account the lethal effects of irradiation. In the experiment of Kim and Evans [9], which will be studied in detail, the doses range from 300 to 1000 R, and the cell death is important at these doses. However, in this study of sublethal effects or irradiation, we shall neglect cells which die. In doing so, we note that the dead cells are included in the experimental data since it is impossible to distinguish dead cells from alive ones. Moreover, some cells which are not immediately killed by irradiation may have in their progeny some cells which die later. Fig. 7 shows the evolution of the mean post-irradiation number of divisions of which nonsurviving cells are capable.

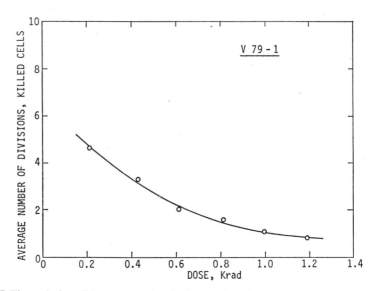

Fig. 7. The evolution of the mean postirradiation number of divisions of which nonsurviving cells are capable, for Chinese hamster cells. (From [4].)

2.2. Simulation

We describe in the appendix a scheme for the mathematical expression of the previous models; even in the simplest cases, such as that of looking only at small times and at asymptotic results, the computations necessary to derive the theoretical formulae giving the percentages of cells in the four phases as a function of the time after irradiation are difficult. On the other hand, our computer method of representation of cells may be easily used to simulate the models with any set of probability functions and without any computing approximation.

Fig. 8 shows the flow chart for the simulation of Model (a), assuming that the block occurs immediately before the transition between G_2 and M. The computations are done on N_{LIM} successive family trees (e.g., $N_{LIM} = 500$). Each family tree is studied during G generations (e.g., $G=7$, and the family tree has $2^7 - 1$ cells). The maximum number of generations we study is 13, this limitation being due to the size of the central memory of the computer UNIVAC 1107. TIR indicates the time of irradiation and is chosen sufficiently large (e.g., equal to the duration of three cell cycles) in order for a simulated population to be considered stationary by this time. For any cell J of the family tree we simply observe if it is irradiated before its transition from G_2 to M. If this is the case, we add the duration of the block (Fig. 4) to $Y(J,3)$ and $Y(J,4)$; this

SUBLETHAL EFFECTS OF IRRADIATION ON CELL KINETICS

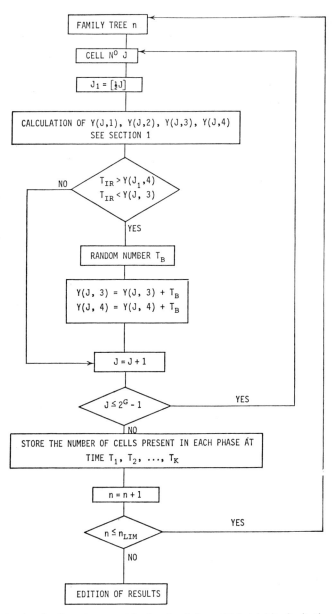

Fig. 8. Example of a flow chart used in the simulation of Model (a) of mitotic delay. T_{IR} indicates the time of the irradiation.

duration is assumed to have a given probability density function. This modification will appear in the family tree; see the recurrence equations (iii) above. When the whole family tree is simulated, the numbers of the cells present in each phase at time $T_1, T_2, \ldots, T_K$ are computed as described above. These results are stored, and a new family tree, if necessary, is simulated, etc.

2.3. Results

2.3.1. Experimental results

We shall attempt to interpret the experimental data given in [9]. In that paper, the authors studied the effects of X-irradiation on the mitotic cycle of Ehrlich ascites tumour cells. As mentioned before, they used three different

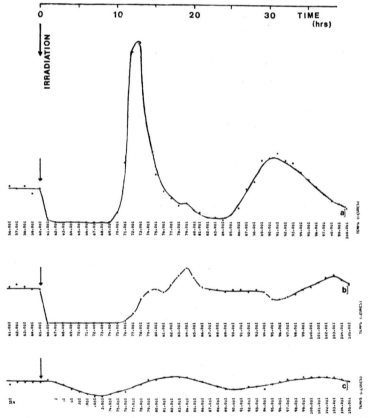

Fig. 9. Typical patterns of evolution of the mitotic index in Models (a), (b) and (c).

doses: 300, 500 and 1000 R. In Fig. 2 were shown the variations of the mitotic index (Fig. 2a) and of the labelling index (Fig. 2b), i.e., the percentage of the cells in the S phase as a function of the time after irradiation. The durations of the phases of the cell cycle are estimated by the authors and will be used in our simulation. The mean values are $G_1 = 5.7$, $S = 8.5$, $G_2 = 3.8$ and $M = 1.0$ h.

2.3.2. *Typical outputs for Models (a), (b) and (c)*

Before giving any interpretation of the experimental curves shown in Fig. 2, we shall describe some typical outputs for Models (a), (b) and (c). For each of these three models, the evolution of the mitotic index is shown in Fig. 9, and the evolution of the percentages of the cells in the three phases are shown in Fig. 10.

In Model (a), we assume that there is a block at the transition between G_2 and M; the duration of this block is set to be 12 h (with a standard deviation of 1.0 h). In Model (b), we assume that the duration of the repair is also equal to 12 h (s.d. = 1.0 h), and that the point R (see Fig. 5) is also the transition $G_2 \rightarrow M$. In Model (c), we assume that the rate of increase of the DNA content

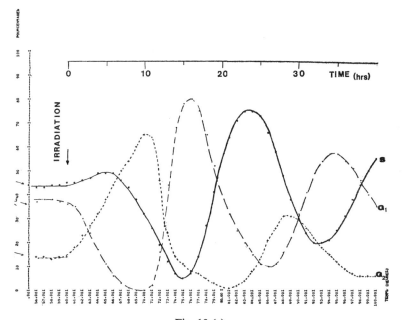

Fig. 10 (a)

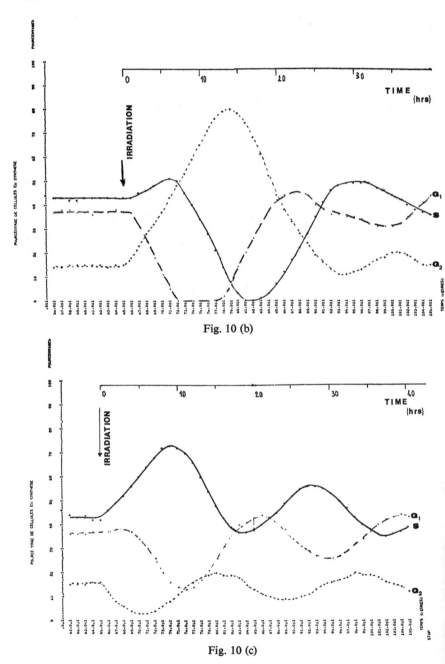

Fig. 10 (b)

Fig. 10 (c)

Fig. 10. Typical evolution of the percentages of cells in the three phases according to Models (a), (b) and (c).

SUBLETHAL EFFECTS OF IRRADIATION ON CELL KINETICS

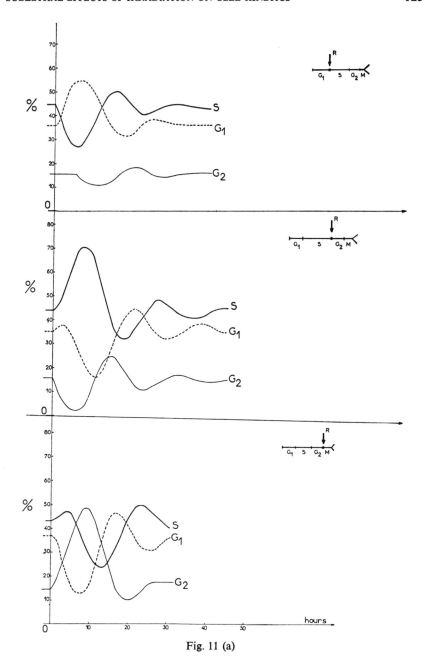

Fig. 11 (a)

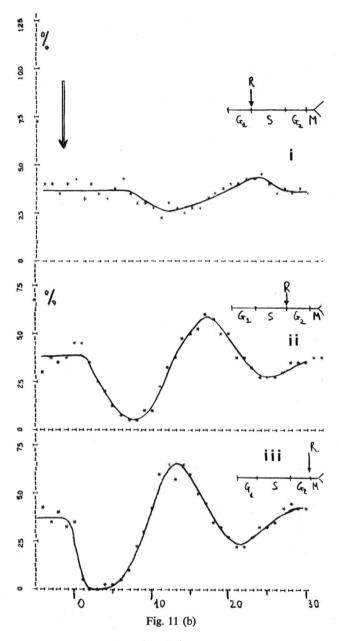

Fig. 11. Evolution of (a) the mitotic index and (b) the percentages of cells in the three phases for different transit points (indicated on the figures),

is, after irradiation, equal to 20% of the initial rate. The difference between the three patterns of evolution of the mitotic index is clear. One can note that in Models (a) and (b) a second peak can be observed after the irradiation at a time approximately equal to the duration of the cell cycle (=19.0 h).

More interesting are the differences observed in Fig. 10 between the three models. For instance, 20 h after the irradiation, in Model (a) we have about 70% of the cells in the S phase, in Model (b) we have only 5% of the cells in the S phase, and in Model (c) we have about 40% of the cells in the S phase. So, if we want to set up a report of the treatment using a drug which kills specifically the S cells, it is important to know which model (if any) is true; the use of this drug in case (a) will be optimal 22 h after the irradiation, and in case (b) 8 h after the irradiation.

In Fig. 11, the influence of the position of the transition point R of Model (b) on the evolution of the mitotic index is shown. In case (i), R is at the transition $G_1 \to S$, in case (ii) R is at the transition $S \to G_2$, in case (iii) R is at the transition $G_2 \to M$. In all three cases we have assumed a mean repair time of 12 h (s.d. = 4.0 h). We see on these theoretical figures that the decrease in the mitotic index indicates the position of the point R in the cell cycle. This result will be used now.

2.3.3. *Fitting of the experimental data*

The results shown in Section 2.3.2 led us to consider only the blocks and repair mechanisms located at the transition $G_2 \to M$ for the three experimental mitotic indexes drop immediately. The mean parameters of the cell cycle, as mentioned before, were given in [9]. We assumed that the durations of the phases were lognormally distributed with the standard deviations 5.0 h for G_1, 4.0 h for S, 2.0 h for G_2 and 0.5 h for M.

In Fig. 12(a) are shown the experimental percentages of cells in the mitotic and synthetic phases after an irradiation of 300 R. The curves correspond to Model (b) (repair) with a time of repair of 12 h (s.d. = 4.0 h). In Fig. 12(b) are shown the data obtained after 500 R; the curves correspond to Model (b) with a time of repair of 17 h (s.d. = 4.0 h). In Fig. 12(c) are shown the data obtained after 1000 R; the curves correspond to Model (b) with a time of repair of 27 h (s.d. = 4.0 h).

We see on these figures that the shapes of the simulated and experimental curves are very close for the three doses. However, in one case (Fig. 12(b), bottom) the experimental curve is bimodal while the theoretical one is not. We have not yet attempted to explain this fact. In all three cases, the repair model provided the most satisfactory fitting for the percentages of the cells both in the mitotic and synthetic phases.

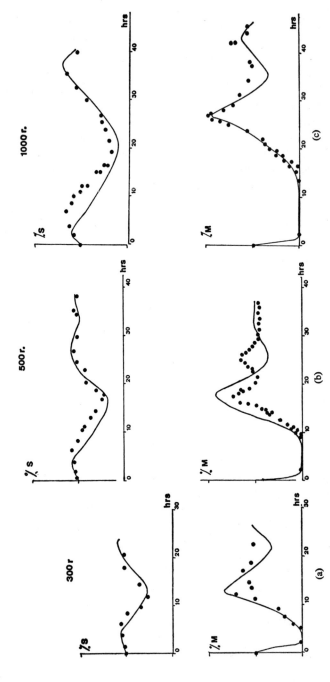

Fig. 12. Experimental points for the percentages of cells in the S and M phases (see Fig. 9) after an irradiation, of (a) 300 r, (b) 500 r and (c) 1000 r; and simulated curves computed by assuming a repair model ((a) $T_R = 12$ h, (b) $T_R = 17$ h, (c) $T_R = 27$ h).

In conclusion, we have shown that the repair model we have set up interprets the experimental data for all three doses studied. It is our aim to check if this model could be applied to other data.

3. Conclusion

Biomathematical methods have been used for many years in the field of cell kinetics. The first approaches were probabilistic and use mainly branching processes ([2, 15], see also the bibliography in [7]). A compartmental approach has been used also by some authors [6, 12, 8]. A purely computer approach has been used by Barrett [1].

The importance of our computer method is in its generality and simplicity i.e., it makes it possible to study various aspects in the field of cell kinetics. We get a close simulation of the reality and not a set of equations which cannot be discussed by most experimentalists.

In addition, the exact mathematical resolution of many cell kinetics problems are difficult and need a vast knowledge of numerical analysis; e.g., in the paper which describes our method [17] we have shown that our method may be used to study the evolution of the histograms of grains after labelling in different subpopulations of cells; even the equations describing these evolutions would be difficult to set.

The limitations of our method are due to the large storage required in the central memory of the computer:

(a) It is difficult to study directly the interactions between cells belonging to the same family tree or belonging to different family trees.

(b) It is difficult to simulate jointly several cell populations and their interactions.

We had some interesting "experiences" in our laboratory, where the same problem has been treated both by a student using our simulation technique and by a student using a biomathematical approach: each time, the former found the result faster than the latter because he modified only a few instructions to the general program and almost immediately began to interpret his data.

The problem of the cost of the simulation techniques is very often posed. Most of the time, authors speak only about computer time; they do not speak about research time and maybe also about the cost for a biological laboratory to get a well-trained mathematician in probability theory branching processes and numerical analysis.

Appendix. Scheme of mathematical resolution

A.1. *The general integral equation*

Let $t = 0$ be the time of irradiation. We shall give the general integral equation describing the evolution of the progeny of a cell present at this time.

Definitions.

g is the probability density function (p.d.f.) and G the cumulative distribution function (c.d.f.) of the duration of the cell cycle (when there is no irradiation).

b and B are the p.d.f. and c.d.f. of the time of division of the first cell (if there is irradiation at $t = 0$); b_1 and B_1 are the p.d.f. and c.d.f. of the time of division of the first cell if there was no irradiation at $t = 0$.

$M(t)$ is the number of cells of the progeny of the first one which are present at time t.

$MI(t)$ is the percentage of cells of the progeny which are in mitosis at time t; T_m is the duration of the mitosis.

With these notations, the following integral equation holds:

$$M(t) = 1 + B(t) - 2G(t) + 2 \int_0^t M(t-u) g(u) \, du. \tag{A.1}$$

The equation (1) may be simply obtained by considering the cells of different age present at time t.

In eq. (1) the model which is used to express the effects of irradiation defines the functions b, B; indeed, in stationary populations of cells, (b_1, B_1) is the asymptotic law of survival time for cells,

$$B_1(t) = a e^{at} \left(1 - 2 \int_0^t e^{-au} g(u) \, du \right),$$

where a is the rate of growth of the population.

Let x be the survival time of a cell when there is no irradiation, y the survival time of a cell after irradiation, $R(x)$ the delay introduced by the irradiation. Then we have $y = x + R(x)$. The law (b_1, B_1) for x being known, it is possible to calculate the law (b, B).

A.2. *Resolution*

Eq. (1) describes the evolution of the number of cells originating from one irradiated at time $t = 0$. In addition, we want to know numerically the evolution of the percentages of cells in the four phases of the cell cycle. We shall describe here only results for large and small values of t.

(i) *For large values of t.* The integral equation described above may be written

$$M = 1 + B - 2G + 2M * g,$$

where * means "product of convolution" or, by using Laplace transforms,

$$\tilde{M}(s) = \frac{1}{s}[1 + \tilde{b}(s) - 2\tilde{g}(s)] + 2\tilde{M}(s)\tilde{g}(s),$$

$$M(s) = \frac{1}{s}\left(\frac{1 + \tilde{b}(s) - 2\tilde{g}(s)}{1 - 2\tilde{g}(s)}\right).$$

If we suppose that $g(t) = \delta_{T_c}$ (no variability), we have immediately the asymptotic solution for large values of t

$$M(t) = Ke^{at},$$

with $\tilde{g}(a) = \frac{1}{2}$ and $a = t_c^{-1}\log 2$ and $K = (\log 2)^{-1}\tilde{b}(a)$.

(ii) *For small values of t.* The integral equation may be written

$$(1 - 2g) * M = 1 + B - 2G.$$

In the algebra of convolution, we have

$$(1 - 2g)^{-1} = 1 + 2g + \ldots + 2^n g * M^{(n)} + \ldots.$$

Let us suppose that the duration of the mitosis (t_m) is short; we can approximate the mitotic index M.I. by the quantity

$$\text{M.I.} = t_m \frac{M'}{M},$$

where

$$M = (1 - 2g)^{-1} * (1 + B - 2G),$$
$$M = 1 + B - 2G + 2g + 2g * B - 4g * G,$$
$$M' = b - 2g + 2g' + 2g * b - 4g * g,$$

for small values of t ($t \ll T_c$), $g(t)$ and $G(t)$ are small and we have

$$\text{M.I.} \approx t_m \frac{b}{1 + B}.$$

We have to note that such approximations make it possible to compute the mitotic index (M.I.) but not the labelling index because the duration of the synthesis is large (30–70% of the total duration of the cell cycle).

References

[1] J. C. Barrett, Optimized parameters for the mitotic cycle, *Cell Tissue Kinet.* 3 (1970) 349–353.
[2] M. S. Bartlett, Distributions associated with cell populations, *Biometrika* 56 (1969) 391–400.
[3] R. G. Canti and F. G. Spear, Effects of gamma rays on mitosis in vitro, *Proc. Roy. Soc. London* B105 (1929) 93–98.
[4] M. M. Elkind, A. Han and K. W. Volz, Radiation response of mammalian cells grown in culture, IV. Dose dependence of division delay and postirradiation growth of surviving and non-surviving Chinese hamster cells, *J. Natl. Cancer Inst.* 30 (1963) 705–721.
[5] E. Frindel and M. Tubiana, Radiobiology and the cell cycle, in: R. Baserga, ed., *The Cell Cycle and Cancer* (Dekker, New York, 1971).
[6] G. M. Hahn, A formalism describing the kinetics of some mammalian cell populations, *Math. Biosci.* 6 (1970) 295–304.
[7] T. E. Harris, *The Theory of Branching Processes* (Springer, Berlin, 1963).
[8] B. Jansson, Mathematical models in cell cycle kinetics, in: N. T. J. Bailey et al., eds., *Mathematical Models in Biology and Medicine* (North-Holland, Amsterdam, 1974) 21–39 (this volume).
[9] J. H. Kim and T. C. Evans, Effects of X-irradiation on the mitotic cycle of Ehrlich ascites tumor cells, *Radiation Res.* 21 (1964) 129–143.
[10] H. E. Kubitschek, The distribution of cell generation times, *Cell Tissue Kinet.* 4(2) (March 1971) 113.
[11] W. K. Sinclair, The combined effect of hydroxyurea and X-ray on Chinese hamster cells in vitro, *Cancer Res.* 28(1968) 198.
[12] M. Takahashi, Theoretical basis for cell cycle analysis, I. Labelled mitosis wave method, *J. Theoret. Biol.* 13(1966) 202–211.
[13] M. Takahashi and K. Inouye, Characteristics of cube root growth of transplantable tumors, *J. Theoret. Biol.* 14(1967) 175–283.
[14] M. Takahashi, Theoretical basis for cell cycle analysis, II. Further studies on labelled mitosis wave method, *J. Theoret. Biol.* 18(1968) 195–209.
[15] E. Trucco and P. J. Brockwell, Percent labelled mitosis curves in exponentially growing cell populations, *J. Theoret. Biol.* 20(1968) 231.
[16] A.-J. Valleron and E. Frindel, Biomodalité ou unimodalité de la durée de G_1. Etude par voie de simulation sur ordinateur, *C.R. Acad. Sci. Paris* D (1970) 271–824.
[17] A.-J. Valleron and E. Frindel, Computer simulation of growing cell populations, *Cell Tissue Kinet.* 1(1973) 69.

COMMENTS

Groth:

Concerning the last three slides you showed us, presenting a comparison between simulated curves and experimental data, I want to ask you what you mean by a "very good fit"? Do you mean from a qualitative or a quantitative point of view?

Furthermore, I think you have given a good example of the power of simulation procedures when analytical methods are laborious to apply.

Valleron (Added in proof):

When I say that the fit between experimental and model curves is good, it is not actually from a quantitative point of view because I have not defined any criteria of distance between experimental and theoretical curves nor attempted to use any automatic technique to get the "best parameters" of this model. But the fit appeared to me to be good in so far as the model used is a simple one (only a few parameters) whereas the experimental data are extremely complex (2 curves at three different doses). I was just entering this subject and I judge it more important at this stage to aim for simplicity.

PANEL DISCUSSION ON THE PHILOSOPHIC ASPECTS OF MATHEMATICAL MODELS IN BIOLOGY AND MEDICINE

PANEL DISCUSSION

QUALITATIVE AND QUANTITATIVE MODELS

R. THOM

Institut des Hautes Etudes Scientifiques, Bures-sur-Yvette, France

It is a generally accepted idea that the main function of the central nervous system of humans and animals is to furnish a simulating copy of the space surrounding the organism and the position of the organism in this space. In general, thanks to the remarkable precision of this copy, the organism is capable of resolving rapidly and effectively the mechanical problems caused by its displacement. In principle this requires solving quantitative problems (to catch a prey, for example), though very often the displacement is due to a choice, to a global strategy of the qualitative nature. In the case of humans, the existence of language shows a great capacity to simulate qualitatively the external processes (physical or biological). Language works as a sensory relay: if an external fact can be observed by individual A, and not by individual B, A can describe orally to B what he observes, and realizing these words, B can imagine qualitatively what happens. So that "modelling" can be considered as a main function of the human psyche either in a qualitative (verbal) or a quantitative form. The scientific "modelling" is only an extension, a refinement of this first function of the mind.

However, the qualitative (verbal) simulation quickly becomes insufficient when it concerns the solving of dynamic situations with an uncertain result, like throwing a projectile at a prey. In this way, the quantitative models of mechanics and physics, which represent the basis of modern science, are found. As a natural consequence one has a tendency to conceive only quantitative models, for the qualitative is nothing but a rough approximation: *qualitative is nothing but poor quantitative*, according to Rutherford's words. Still one must not exaggerate the ampleness of the domain covered by quantitative models. In fact, every quantitative model appeals either to the "unreasonable" exactitude of the fundamental physical laws (according to E. Wigner's expression), or, when it concerns discrete events, to the considerations of statistics (as the law of large numbers). The latter case does not permit, in principle, prevision of the individual event. The former case is not of a validity as wide as one usually believes. In fact, a quantitative law has to possess an expression, independent of the units with which one measures the physical values, appearing in the equation. This imposes a heavy constraint upon the mathematical form of the law (the constraint of dimensionality), and this shows that in a certain sense the quantitative law is linked with the affine group of the dilatations of space–time. Therefore, in fact, it is a consequence of the geometry of space–time. That a phenomenon be ruled by a quantitative model amounts to the demand that it be reducible to the geometry of space–time. Though a professional geometrician, I do not find this exigency reasonable at all.

If, in the studied phenomena, one can only isolate partial processes ruled by the known physical laws, one can only hope to be able to write some quantitative relations; one must complete the model with hypotheses – possibly the most natural ones. Well, the confrontation of the model with the experimental data very often will be done on a qualitative basis. The study of the qualitative properties of the functions (growth, decrease, existence of points of inflection etc.) in itself will be able to give the best confirmation (or rejection) of the proposed model. One can see now why the affirmation of Rutherford is erroneous. The qualitative now is accessible to the mathematicians. It is expressed in the topological properties of the graphs of the relations given by the model. The interpretation of the morphology of the empiric graphs, such as that of the "clouds of points" given by the statistical study of a phenomenon, is a preliminary stage of all modelling. There is no doubt about it – the problem concerned is a difficult one, and its investigation has hardly begun. But one cannot do without a qualitative analysis, the usual language of which shows a character both indispensable and effective.

PANEL DISCUSSION

ADAPTIVE MODEL BUILDING AND COMPUTER ASSISTED ANALYSIS OF BIOMEDICAL DATA

T. GROTH

Uppsala Data Centre, University of Uppsala, Uppsala, Sweden

Computer modelling and simulation has evolved in parallel with modern computers and has become a powerful technique with great potential in all disciplines. Its greatest impact has been in the physical sciences, but the technique can be applied to virtually every discipline in which phenomena can be quantified and represented by symbolic models. The ability of the computer to accumulate and reduce data has been of primary interest to experimental scientists, to which the promise of the computer is a relief from a growing burden of statistical work and other non-creative tasks that increasingly are reducing their effectiveness as scientists. The most important role of the computer in science is, however, as a modelling device and as such it can be used advantageously in the cyclic adaptive interaction between experimental and theoretical work:

(1) design of experiments;
(2) generation of data by laboratory experiments;
(3) reduction of data;
(4) formulation or modification of a model;
(5) conformation of the model to data and judgment of the model;
(6) simulation of new conditions, etc.

The theoretical development in every science is based on the modelling of processes. The physical sciences have historically used various types of mathematical models in their theory [13, 14, 15], and during the last decade, the physical "way of model-thinking" has increasingly influenced the activity in various interdisciplinary fields of research, e.g., biomedicine and biomathematics. The mathematical model is basically meant to serve as a calculating device from which answers can be computed to any questions about the behaviour of the modelled system. Using conventional methods of calculation, one mostly has to sacrifice complexity and accuracy for simplicity and approximation in order to produce working hypotheses. The computer, however, permits the treatment of very complex systems with a large number of variables. This is of special importance in systems analysis and operations research. In problems where the solution cannot be given in a closed analytical form or by numerical methods, the process can often be successfully modelled in the computer by a direct imitation of the natural events in small time steps. This procedure, usually called simulation, is quite different from any conventional form of hand computation. Various probability distributions, produced by random number generators in the computer, are then often useful, especially in certain types of stochastic simulations.

The complexity of biological systems often make them difficult to describe in terms of physical models. Until the nature of the components of the system is known, one has to describe them by so-called "lumped parameters" or by various empiric–stochastic functions. The analysis of tracer data by compartmental models and the description of intravascular transport, and transport of diffusable substances through capillary networks by stochastic methods [4] are some examples. The primary goal of these approaches may often be to derive estimates of hard-to-measure variables, useful for interpretative purposes, from variables that can be measured directly. However, in order to provide a basis for the more distant goal of describing the system by a physical model, the model concepts must be chosen in the physical frame of reference (for an example see e.g. [7, 9]).

The advantages of using a physical model as a link between the system of interest and the mathematical model describing it should be stressed. Compared to empirical model building, the procedure of physical model building is more restrictive in the selection of mathematical representation, and thus gives a larger decrease in inductive entropy (see below). Since the formulation of the model can be effectively based on past experience, the separation of noise from significant information in the experimental data can be done less arbitrarily. Used in this way, physical models are of great interest not only within special fields of physics, but also in other disciplines within the natural sciences.

In order to perform a meaningful model study, the outcome from the model must be checked against experimental findings about the studied system. This is usually done by least-square analysis giving "best" estimates of the model parameters. Non-linear parameter search methods [16] are then necessary tools as well as statistical methods for the estimation of confidence regions of the parameter estimates [5]. In this connection the importance of a complete experimental range of observation should be stressed (see e.g.

[1, 10]), and also the possibility of designing optimal experiments by simulation of various alternative measuring procedures.

Because of the limited accuracy in all measurements, one is often faced with the problem of having several models of entirely different structure giving equal predictions within experimental errors. In problems where the compatibility to data is of main interest, any of the models would serve equally well as a calculating device and the choice between them must be made on other grounds, e.g. ease of calculation or simplicity in the model formulation. When the purpose of the model is to "explain" the mechanisms of certain processes, the question of uniqueness becomes more urgent. However, no model studies will result in a completely unique model. Once a compatible model has been found, one can find infinitely many more complex models, which will agree with the data at least as well as the chosen model. Thus, it has no meaning to talk about "true" models in the classical sense, a fact that has been obvious to physicists for a long time. The choice between models, as "explanations" of processes, must be made on the basis of the degree of preferential confidence or credibility [18], as obtained from previous knowledge. In order to increase the adequacy [6] or confirmability [18] of a model and make it an efficient information processing tool, one has to adjust the complexity of the model to the requirements of all available, relevant, experimental data and at the same time minimize complexity so that the parameters of interest are well determined by the data (see e.g. [1, 2]). The "art of model building" is thus to make a proper lumping of the vast number of parameters on various levels of resolution of a complex system and get a smaller, optimal number of parameters which are defined in a meaningful way in terms of measurable quantities.

Model building is not an endpoint in itself, but should be regarded as a useful logical tool for testing the consistency of theoretical concepts and as a means of designing critical experiments The scientific emphasis is not on the question of truth or uniqueness of the model, but on the clarity with which the discrepancies between the model predictions and experimental data lead to new and profitable interferences. (For illustrative examples see [1, 3].) From the information-theoretical point of view, the gradual concentration and convergence on fewer and fewer hypotheses may be regarded as a decrease in inductive entropy, as defined by the distribution of confidence on competitive hypotheses (the "inverse H-theorem" [18]). Whether or not this may be regarded as a "production of information" is an interesting problem, which has been discussed to some extent by Watanabe [19] and Ruyer [17].

The predictive potential of good models may be used advantageously in a systematic investigation of the performance of a complex system (systems analysis) the results of which can be used in the design or in the optimization of the system. The general "philosophy" of systems analysis is applicable to a wide range of problems (cf. [8, 11, 12]).

References

[1] G. Arturson, T. Groth and G. Grotte, Human glomerular membrane porosity and filtration pressure. Dextran clearance data analysed by theoretical models, *Clin. Sci.* 40 (1971) 137.
[2] G. Arturson, T. Groth and G. Grotte, The functional ultrastructure of the blood-lymph barrier. Computer analysis of data from dog heart-lymph experiments using theoretical models, *Acta Physiol. Scand. Suppl.* 374 (1972) 30 pp.
[3] G. Arturson, T. Groth, G. Grotte, P. Malmberg, R. Samuelsson and U. Sjöstrand, The adequacy and compatibility of compartmental models of electrolyte exchange in the dog's heart, *Uppsala J. Med. Sci.* 78(1973) 169.
[4] J. B. Bassingthwaighte, *Science* 167 (1970) 1347.
[5] E. M. L. Beale, Confidence regions in non-linear estimation, *J. Roy. Statist. Soc.* B22 (1960) 41.
[6] G. E. P. Box, Fitting empirical data, *Ann. N. Y. Acad. Sci.* 86 (1960) 792.
[7] L. Garby and T. Groth, Determination of kinetic parameters of platelet survival by computer simulation, in: J. M. Paulus, ed., *Platelet Kinetics* (North-Holland, Amsterdam, 1971).
[8] L. Garby and T. Groth, The interpretation of red cell survival curves in non-stationary states, *Scand. J. Haematol.* 7 (1970).
[9] L. Garby, T. Groth and W. Schneider, Determination of kinetic parameters of red blood cell survival by computer simulation, *Computers and Biomed. Res.* 2 (1969) 229.
[10] T. Groth and B. Tengström, Clearance of I^{125} sodium iothalamate. Analysis of data from single injection experiments, to appear.
[11] T. Groth and L.-E. Bratteby, Blood volume estimation in hemorrhage and shock. A computer simulation approach, in *Proc. Journées d'Informatique Médicale, Toulouse* (1973).
[12] S. Graffman, T. Groth, B. Jung, G. Sköllermo and J. E. Snell, On the optimisation of dose distribution in radiotherapy. A systems analysis of the influence of various sources of error and uncertainty, in: *Proc 4th Internl. Conf. on Computers in Radiation Therapy* (Uppsala, 1972).
[13] F. Hund, Denkschemata und Modelle in der Physik, *Studium Generale* 3 (1965) 174.
[14] M. Jammer, Die Entwicklung des Modellbegriffes in den physikalischen Wissenschaften, *Studium Generale* 3 (1965) 166.
[15] M. Kac, *Science* 166 (1969) 695.
[16] M. J. D. Powell, in: J. Walsh, ed., *Numerical Analysis – An Introduction* (Academic Press, New York, 1966).
[17] R. Ruyer, *La Cybernétique et l'Origine de l'Information* (Flammarion, Paris, 1954).
[18] S. Watanabe, Information-theoretical aspects of inductive and deductive inference, *IBM J. Res. Develop.* 4 (1960) 208.
[19] S. Watanabe, The cybernetic view of time, in: N. Wiener and J. P. Schadé, eds., *Progress in Biocybernetics*, Vol. 3 (Elsevier, Amsterdam, 1966).

PANEL DISCUSSION

MODELLING

N. T. J. BAILEY

World Health Organization, Geneva, Switzerland

May I first thank the two opening speakers for their prepared contributions, which emphasize some major problems in the use of mathematical modelling in biology and medicine. Although I think that it is a good idea to relate the general philosophical aspects of modelling to some concrete practical problems, we do not, unfortunately, have time today to investigate any individual case histories in very great detail. On the other hand, it would be undesirable to confine our exchange of views on modelling to purely theoretical considerations. I hope that, in the general discussion that we are going to have, it will be possible to achieve some kind of compromise, relating theoretical and philosophical aspects to their practical implications.

Before inviting comments and contributions from all of those present who might wish to do so, may I briefly re-emphasize some of the major points of difficulty that have already been referred to.

There is always a primary problem of deciding how complex a model must be to achieve a reasonable degree of realism, as compared with the level of simplicity required to ensure tractability and ease of interpretation. How should we decide on the correct balance? Again, some aspects of models have to be taken for granted, but others are more in the nature of hypotheses and should be tested. Thus some forms of evaluation are related mainly to an analysis and interpretation of logical structure, the emphasis being on logical and mathematical analysis. Other forms of evaluation are inevitably involved in empirical testing, the carrying out of suitably designed experiments or carefully chosen observations. Finally, in many kinds of systems models we can only proceed by deliberately ignoring much of the fine structure of which we are well aware. The right sort of sensitivity analysis may help to determine whether the results obtained are likely to be much affected by appreciable variations in the assumptions.

I should now be glad to ask for any comments, suggestions or contributions, that anyone may care to make, to the general philosophy of the use of mathematical models in biology or medicine. And I hope that the range of ideas already expressed in the opening will be sufficient to get us started.

PANEL DISCUSSION

PRAGMATIC JUSTIFICATION OF THE "MODEL" IN MEDICINE

J. MARTIN

Section d'Informatique Médicale, Faculté de Médicine, Université de Nancy, Nancy, France

and

C. MONOT

Centre National de la Recherche Scientifique, Nancy, France

We should like to present a very different point of view, that of the user, for whom it is not a question of explaining the reality, but of representing it so as to understand its properties better, analyse them better and make use of them in the medical profession. For this purpose it is necessary to take up the argument from the beginning and ask what is the use of this tool known as "modelling" ("modélisation" in French) and by what method it may be utilized.

1. Evolutive systems and some definitions

The biologist faced with phenomena which he wishes to observe may be compared with the electronics specialist in front of what he calls a "black box". This expression means a "system" in which entries and exits are measurable quantities but whose internal components are assumed to be inaccessible. From these measurements, a model is sought which would enable forecasting the later "evolution" of the "system".

In this way, when a biologist studies phenomena of exchanges or metabolisms, he inspects the doses injected and measures certain concentrations and certain eliminations. Generally, not all stages of the process are accessible. To forecast internal behaviour he is obliged to make hypotheses.

The abstract study of the "evolutive systems" has been pursued and the logical analysis of the various possible structures has enabled classification [8]. This classification is principally based on the role of the time variable. An evolutive system may depend on absolute time. This is the case for obsolescence: the system is known as *extravert*. What follows will be different according to the moment at which the observation started. In other simpler cases, the evolution can be predicted whatever the initial moment: the system is *introvert*.

Two very important categories must also be considered according to whether the system is one with "memory" or otherwise. Take, for example, the kinetics of a substance absorbed in small quantities at the moment t_0 by an organism in a state of equilibrium. The kinetics depend only on the "internal time" $t - t_0$. The system is then called *transitive*. On the contrary, if the organism is not in a state of equilibrium, its evolution will depend on its past (stocking or deficiency, etc.). The system is *hereditary* or *with memory*. Fortunately, in many cases there is a "forgetting function", and only the recent past plays a role.

The advantage of this abstract analysis is to make us understand better the implicit hypotheses during the construction of a logical model or mathematical model of a biological phenomenon. Indeed it is very generally supposed in biology, particularly in metabolic models, that the system is introvert and without memory.

2. Methodology

2.1. *Analysis of induced reactions*

Let us go back to our comparison of the behaviour of the biologist and the electronics specialist.

The latter possesses various approaches enabling him to know the reactions of the device being studied. These approaches are represented by the functions shown in Fig. 1. The function of Fig. 1(a) is often replaced by a series of periodic sinusoidal functions. The quantities of the exits are then studied as a function of size, phase, harmonics, etc. In theory, the first two functions (impulse and degree) enable all the responses of a linear system to be predicted.

The biologist (or the physician carrying out the functional exploration) does not have that many possibilities. Except for functional pulmonary exploration, it is difficult to call on periodically variable entries. When the

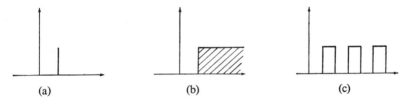

Fig. 1. Functions representing induced reactions. (a) Impulse (Dirac function); (b) degree; (c) battlements.

metabolism of a compound which can be followed in the organism (by labelling or another method) is concerned, a disturbance may be entered:

(i) an impulse: almost instantaneous injection (intravenous);

(ii) a degree: continuous perfusion.

The concentration of the compound is then studied at various accessible stages of its metabolism. It should be remarked, in fact, that the "black box" can be broken down into several compartments when certain internal quantities can be verified.

2.2. *The techniques of analysis of the results*

2.2.1. *Research for descriptive indices*

The physician is generally satisfied with the morphological description of the exit curves: moment and maximum rate, etc. The statistical correlation of the morphological criteria is then sought for, with the clinical elements of diagnosis or prognosis. Thus a radioactive iodine fixation rate exceeding 45% in 24 h indicates thyroid dysfunctioning.

2.2.2. *Comparison of experimental curves*

Without any preliminary hypothesis and when there are evolutive curves corresponding to various stages of the process being studied (we will call them $f(t)$ and $g(t)$), exact mathematical methods, utilizing the "convolution product" of the two functions, permit the definition of an index connected with time. One word should be said as to the principle. For different values of T, the following function is formed:

$$C(T) = \int_0^\infty f(t)\, g(t - T)\, dt.$$

$C(T)$ generally goes through a maximum for a particular value of T, which is taken as an index.

In this way Suquet [7] (see also [3]), studying the rate of labelled ionic iodine ($f(t)$) and the rate of the various labelled organic fractions ($g(t)$) in the blood (see Fig. 2), shows that this provides a criterion giving better separation

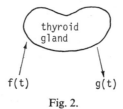

Fig. 2.

of the clinical groups in accordance with the physiopathology than the usual morphological indices.

2.2.3. Construction of a model

Instead of considering phenomena from the exterior as before, an attempt may be made to reconstruct them from within. For this purpose, it is generally necessary to make hypotheses as to the principal stages of the process and formulate them mathematically. Mathematical constraints may require biological reality to conform to a given formal plan. We shall refer to the difficulties therein later.

3. The mathematical model

3.1. The first step: general architecture of the model

The architecture should strive to take into account all the known facts, both physiologically and pathologically. An example is the metabolism of inorganic labelled iodine.

What happens during the hour following injection in the blood (intravenous route: "impulse entry") of a small quantity of inorganic labelled iodine? It spreads throughout the organism (blood and interstitial liquids) S and L. Part of it is absorbed by the thyroid gland (T) to form hormones, another part is absorbed by the digestive tract (G), and a fraction, finally, is eliminated by the kidneys (R). Several methods of representation are possible. The first assumes passage from one box (compartment) to another (see Fig. 3). The second representation puts the accent more on the movements of the product (see Fig. 4). Note that the stages (and their associated "compartments") can be subdivided. There can be "delay" in a compartment (and the corresponding stage). It is also possible, for example, to be entitled to analyse the thyroid gland as being composed of several compartments, etc. In fact, much depends on the truly measurable quantities. The physiological significance of the compartments is complex, and reference should be made to [6] for this analysis.

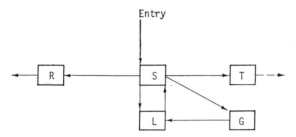

Fig. 3. Compartment representation of the spreading of inorganic labelled iodine throughout the organism. S – blood; L – interstitial liquids; T – thyroid gland; G – digestive tract; R – kidneys.

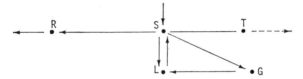

Fig. 4. Movement representation of the spreading of inorganic labelled iodine throughout the organism. Legend: see Fig. 3.

3.2. *Mathematical formulation*

The diagrams shown in Figs. 3 and 4 are purely qualitative. To arrive at a utilizable formula, hypotheses of the movements must be made (exchanges, transfer kinetics) (for the biological applications see [5, 6]).

According to the case, different types of differential equation systems are arrived at, as we indicated in the work which we presented [4]. Under certain somewhat restrictive hypotheses, these systems have constant coefficients. However, a good number of biological phenomena do not follow linear laws. Enzymic kinetics are not systematic, and there may also be nonlinear rules, etc. This is probably the case for the phenomenon of hyperglycemia induced by an intravenous injection. In fact, this is a case in which the coefficients of passage from one compartment to another are regulated by the very quantity they are controlling [1].

4. Utilization of the mathematical model

4.1. *The control of the formulation*

A first period of "simulation" (analogical or numerical) shows whether the proposed model allows experimental curves to be found or not. In certain cases this enables certain hypotheses to be eliminated when they are too complex for their incompatibility to be shown by formal examination. This is what occurred to the usual clinical interpretation of the isotope nephrogram [2].

Unfortunately, a correct reconstruction of the experimental results does not guarantee the validity of the model. It is a necessary but absolutely insufficient condition. Only the soundness of the facts used for its construction represents a good pledge for conformity.

If the model is a good reflection of reality, certain pathological phenomena can be forecast or sometimes explained by varying the parameters of the model. This means that one has a remarkable tool for research and analysis.

Furthermore, this raises a very interesting philosophical problem: are pathological conditions representable by extreme values of normal parameters, or do they call on new mechanisms (or normally negligible mechanisms)? We would emphasize the fact that this is solely a representation. However, we consider that a model which takes various conditions into account is plausible, at least partially.

4.2. *The resolution of the inverse problem*

If the proposed model is accepted – at least subject to inventory – what we consider the real fundamental work may be started: the resolution of the inverse problem. This means determining the values of the parameters of the model which best account for a given network of experimental curves. To do so, it is often necessary to adapt already known methods, and even sometimes to perfect new methods for the resolution.

4.3. *A few difficulties*

4.3.1. *The experimental data*
In medicine, experimental curves are often known only by certain points. Moreover, these points, limited in number, are poorly distributed with regard to the "accidents" of the curve. Thus the moment of exact maximum of radioactive iodine fixation by the thyroid is unknown. In practice, the measurement

at 24 hours is taken as the value of this maximum, which is sometimes completely incorrect.

The lack of precision of certain data should also be noted: All measurements of radioactivity follow a Poisson law; the measuring apparatus may show a deviation, etc.

4.3.2. *The structure of the model is quite difficult to obtain*

Simplifications in drawing up the model may be a reason for failure. In this case we are often tempted to make a new model of more complex type, but then the number of unknown factors increases, while the number of known experimental curves remains the same. Mathematical and technical difficulties are also added:

(i) resolution of large systems;

(ii) the risk of approaching the solution by several different groups of values.

4.3.3. *Basic hypotheses*

Hypotheses are to be made to enable the construction of a system with compartments which must be verified as far as the problem allows. We would mention:

(i) linearity of exchanges; we have referred to this already;

(ii) dilutions and mixtures are not always obtained instantaneously (an example is the isotope nephrogram);

(iii) the notion of "compartment" may be purely statistical, as in the case of a group of cells serving the same capacity;

etc.

5. Conclusion

We have gone through the various stages of the application of mathematics in the study of the phenomena of biological exchanges, with the study of metabolisms as a particular case. These stages are:

(1) The definition of "system" and as strict as possible determination of the measurable "causes", that is, the parameters which will be taken into account.

(2) Formulation of the model by carefully specifying the hypotheses and the experimental data available (examples are instantaneousness of mixtures, etc.). This architecture should be the best possible reflection of all known facts.

(3) Mathematical formulation follows the preceding analysis, but brings with it new constraints. It often requires hypotheses, such as the linearity of exchanges, which must be verified.

(4) Use of the model:
  (a) simulation; with a study of the influence of the various parameters, qualitative approach to the curves;
  (b) resolution of the inverse problem; search for characteristic qualities such as regulatory functions, exchange coefficients, etc.

When this work has been carried out, it will be found to provide an irreplaceable tool. It demonstrates the incompatibility of certain hypotheses and reveals our ignorance. From the practical point of view, when a good model has been found it enables the interpretation of the experimental curves to be based no longer on a few morphological accidents but on their underlying causes. The biologist can then be of real assistance to the clinician. It is a good start on the road to true quantitative biology.

This utilization of the "model" is different from that propounded at the beginning of this round-table conference, but it is not inconsistent with it. In the history of sciences, general theories have always preceded and followed particular experiments. This double approach is fundamental for a good understanding of reality.

References

[1] O. Foucaut and J. Martin, Non linéarité de certains modèles compartimentaux. La simulation numérique de l'hyperglycémie provoquée par voie intraveineuse, *Medical Information Days, Toulouse* (Muray, Paris, 1969).

[2] J. Martin and C. Monot, Analyse d'un modèle mathématique du fonctionnement rénal. Ajustement numérique sur ordinateur des paramètres à partir des courbes du néphrogramme isotopique à I^{131} (Hippuran), in: L. Timmermans and G. Merchie, eds., *Radioisotopes in the Diagnosis of Diseases of the Kidneys and the Urinary Tract* (Excerpta Medica, Amsterdam, 1969) 508–520.

[3] E. Mitsoras, Development of a new methodology for functional exploration, Thesis, University of Montpellier (1967).

[4] C. Monot and J. Martin, Reflections on some algorithms for the solution of the inverse problem (identification or adjustment) for linear compartmental models, in: N. T. J. Bailey et al., eds., *Mathematical Models in Biology and Medicine* (North-Holland, Amsterdam, 1974) 49–70 (this volume).

[5] A. Rescigno and G. Segre, *Drug and Tracer Kinetics* (Blaisdell, Waltham, Mass., 1966).

[6] C. W. Sheppard, *Basic Principles of the Tracer Method* (Wiley, New York, 1962).

[7] P. Suquet, Personal communication (Montpellier).

[8] I. Vogel, *Théorie des Systèmes Evolutifs* (Gauthier-Villars, Paris, 1966).

AUTHOR INDEX

d'Arcy Thompson, 93, 95
Arturson, G., 66, 140

Bachev, S., 1–7
Bäckman, G., 97–110
Bailey, N. T. J., 9–11, 130, 141–142, 150
Barrett, J. C., 23, 38, 127, 130
Bartlett, M. S., 130
Baserga, R., 39, 130
Bassingthwaighte, J. B., 140
Beale, E. M. L., 140
Beaugas, G., 66
Belin, J. P., 66
Berdot, P., 65
Bietry, J. C., 65
Borlaug, F., 72
Box, G. E. P., 140
Bratteby, L.-E., 140
Brauner, J. C., 48
Brockwell, P. J., 130
Browning, H., 72

Canti, R. G., 113, 130
Cherruault, Y., 65

Delbrück, M., 93, 95
Dresch, C., 66
Dubus, D., 48

Elkind, M. M., 130
Evans, T. C., 112, 117, 130

Faille, A., 66
Falck, M., 78
Faurré, P., 65
Feldstein, M. S., 100, 110
Finas, J., 65
Forrester, J., 10
Foucaut, O., 65, 150
Frey, T., 72

Frindel, E., 113, 130
Fry, R. J. M., 39

Garby, L., 140
Georges, F., 65
Gibler, J., 72
Graffman, S., 140
Groth, T., 66, 67, 68, 78, 130, 137–140
Grotte, G.. 140

Hahn, G. M., 130
Han, A., 130
Hanes, S., 23, 39
Härö, A. S., 97–110
Harris, T. E., 130
Hempel, G., 13–20
Hentsch, W., 66
Hogg, J. D., 23, 39
Howard, A., 21, 38
Hund, F., 140

Inouye, K., 130
Irving, E., 65

Jammer, M., 140
Jansson, B., 21–39, 130
Joly, G., 48
Jung, B., 140

Kac, M., 140
Kalimo, E., 110
Kernevez, J. P., 41–48
Kim, J. H., 112, 117, 130
Korobelnik, G., 65
Kubitschek, H. E., 130

Lamerton, L. F., 39
Legras, J., 65
Lions, J. L., 48
Loridan, P., 65

Malmberg, P., 140
Martin, J., 49–70, 143–150
Mary, J. Y., 69
Mateeva, E., 91
Mather, J., 94, 95
Mattila, A., 110
Meadows, D., 10
Mendelsohn, M. L., 23, 31, 39
Merchie, G., 65
Mitsoras, E., 150
Monot, C., 49–70, 143–150

Najean, Y., 66
Nissen, G., 65

Pagurova, W. J., 18, 20
Parsons, D. H., 66
Paulus, J. M., 140
Pelc, S. R., 21, 38
Penel, P., 48
Perälä, J., 97–110
Pešek, J., 71–76
Petkova, E., 1–7
Powell, M. J. D., 67, 140
Putt, H., 72

Quastler, H., 22, 39

Reich, J. G., 48, 68, 77–79
Rescigno, A., 150
Révész, L., 24, 39
Rhode, K., 78
Robert, J., 65
Rubinow, S., 66
Ruyer, R., 139, 140

Samuelsson, R., 140
Schadé, J. P., 140
Schneider, W., 140
Schwartz, L., 66
Segre, G., 150

Sendov, Bl., 81–92
Sheppard, C. W., 66, 150
Sinclair, W. K., 116, 130
Sjöstrand, U., 140
Sköllermo, G., 140
Snell, J. E., 140
Spear, F. G., 113, 130
Steel, G. G., 23, 39
Suneson, E., 72
Suquet, P., 146, 150

Takahashi, M., 23, 24, 31, 39, 130
Tannock, I. F., 24, 25, 39
Tengström, G., 67, 140
Thom, R., 93–95, 135–136
Thomas, D., 48
Timmermans, L., 65
Tonnellier, F., 65
Trucco, E., 130
Tsanev, R., 81–92
Tubiana, M., 113, 130

Väänänen, I., 97–110
Vallée, G., 66
Valleron, A.-J., 67, 69, 111–131
Vauhkonen, O., 97–110
Vogel, I., 150
Volz, K. W., 130
Voss, K., 13, 20

Waddington, C. H., 93, 95
Walsh, J., 140
Wangermann, G., 78
Watanabe, S., 139, 140
Wiener, N., 140
Wigner, E., 135
Winzer, A., 66

Yvon, J. P., 48

Zinke, I., 77–79